Planning Science Instruction for Emergent Bilinguals

Planning Science Instruction for Emergent Bilinguals

Weaving in Rich and Relevant Language Support

Edward G. Lyon and Kelly M. Mackura

Foreword by George C. Bunch

TEACHERS COLLEGE PRESS

TEACHERS COLLEGE | COLUMBIA UNIVERSITY
NEW YORK AND LONDON

Published by Teachers College Press,® 1234 Amsterdam Avenue, New York, NY 10027

Copyright © 2023 by Teachers College, Columbia University

Front cover by Rebecca Lown Design.

Library of Congress Cataloging-in-Publication Data is available at loc.gov

ISBN 978-0-8077-6808-2 (paper)
ISBN 978-0-8077-6809-9 (hardcover)
ISBN 978-0-8077-8159-3 (ebook)

Printed on acid-free paper
Manufactured in the United States of America

I dedicate this book to my family—Adelyn, Evan, and Noah—for their ever-present love and support. I also dedicate it to all the preservice teachers and mentors who I have worked with over the years. Finally, I dedicate this book to the two sheltered learning classes I taught as a first-year teacher. You will always keep me grounded in the critical importance of rich and relevant language support.

—Edward G. Lyon

I dedicate this book to my family and tribe—thank you for your continued love and support of this important work. I also dedicate this book to all my fellow educators—thank you for your powerful collaboration and always challenging me to grow in how to best serve students and families. Finally, I dedicate this book to my students and families past, present, and future—thank you for the ongoing privilege of learning and growing with you in community.

—Kelly M. Mackura

Contents

PART III: PLANNING A CONCEPTUAL AND LINGUISTIC PROGRESSION OF LEARNING

Foreword

In my course designed to prepare future secondary content-area teachers for working with multilingual learners, including those classified by their schools as English Learners (ELs), I frequently tell teacher candidates to keep their content-area knowledge, understandings, and passions at the center of their work with this population. I warn these preservice teachers that, throughout their careers, they are likely to be inundated with people and initiatives promising advice or training on how to work with EL-classified students (what the authors call in this book emergent bilinguals). I suggest that they be particularly skeptical of lists of "EL strategies" that can be used in "any subject area." And I urge them to resist approaches that seem to take teachers and their students, in the name of language support, *away* from the disciplinary practices and understandings at the heart of their subject areas.

Now, thanks to Edward Lyon and Kelly Mackura's conceptually rich and highly practical book, teachers—at any stage of their careers—can *see* what it might look like to plan language support for multilingual learners that keeps disciplinary content, concepts, and practices at the heart of science instruction. As the number of students learning English as an additional language grows throughout the United States, it becomes increasingly important to maintain, and, where necessary, enhance, the quality, relevance, and intellectual challenge of disciplinary instruction for this population—rather than diminish these learning opportunities in the name of language "support." What is needed, in other words, is *amplification* of disciplinary learning opportunities for EL-classified students—through enhancement, elaboration, extension, and scaffolding—rather than the *simplification* of both content and language that has all too often been the proposed remedy (Walqui & Bunch, 2019).

Lyon and Mackura, who collectively are experienced classroom science teachers, teacher educators, and science education researchers, and who have worked with multilingual students throughout their careers, provide a clear vision and detailed guidance for planning explicit language support for EL-classified students and other multilingual learners. But it is never in doubt that robust, standards-based science instruction is the anchor for this support.

This orientation is clearly articulated in the arguments that the authors advance in the introductory chapters, but it becomes especially prominent in the actual processes and tools they suggest for teachers to use when designing and planning units. Lyon and Mackura start by showing teachers how they can thoughtfully and substantively consider the knowledge, backgrounds, and interests of their multilingual learners relevant to the target science focus. They next ask teachers to carefully consider the Next Generation Science Standards (NGSS) applicable to their topic of instruction, identify performance expectations, and select a "big idea statement." After this disciplinary foundation for the unit is in place, teachers are encouraged to begin considering the role that language might play in students' engagement with the key ideas and practices and what tools might be helpful for promoting the development of this language.

For example, in the unit on Earth's ecosystems and resources that Lyon and Mackura use throughout the book to illustrate their approach, readers are taken through the selection of key performance expectations based on the three dimensions of the NGSS: disciplinary *core ideas* (in the case of this unit, interdependent relationships in ecosystems), crosscutting *concepts* (cause and effect in this unit), and science and engineering *practices* (e.g., analyzing and interpreting data to provide evidence for a phenomenon). It is this disciplinary grounding that sets the stage for the authors' guidance for how teachers can support EL-classified students in navigating—and further developing—the language associated with the science. The authors suggest using state English language development (ELD) and English language arts (ELA) standards as resources for identifying points of language focus—but only in direct conjunction with the key practices at the heart of the unit. For teachers, this process involves carefully considering the key language functions related to scientific practices themselves and how a focus on language can be woven into the core science instruction.

For example, in the Earth ecosystems unit, students are expected to "construct an explanation that predicts patterns of interactions among organisms across multiple ecosystems." Because "Constructing an Explanation" and "Analyzing and Interpreting Data" are two key science practices necessary to meet this expectation, the language functions *explain* and *analyze* can be viewed as anchors for language focus. The authors show how teachers can use state ELD and ELA standards (California's in this case) to work with students on interpretive, collaborative, and presentational language practices that support these functions.

The authors' approach to vocabulary also demonstrates this discipline-first approach. They recommend that teachers, when planning a unit, consider some of the key content-area vocabulary that will be important for engaging in the scientific practices, but *not* to try to "frontload" it all before the lesson begins. Instead, the authors point out the importance of engaging students first in developing conceptual understanding as students use their own language (their developing English, their home languages, nonscientific language, gestures, or visual representations). Teachers can then, as students are in this midst of this engagement with key ideas, gradually support their development of related scientific vocabulary in order for them to communicate those ideas more clearly and precisely.

Throughout this process, it is important to keep in mind that, as Lyon and Mackura argue, beginning with key disciplinary science ideas and practices as teachers' foundation for integrating language support for EL-classified students not only offers these students more equitable content-area learning opportunities, but also facilitates their language development. The authors in this book have not only argued this point convincingly, but have provided helpful guidance, concrete examples, and customizable tools for teachers to do this challenging and necessary work themselves.

—George C. Bunch,
University of California,
Santa Cruz

REFERENCE

Walqui, A., & Bunch, G. C. (Eds.). (2019). *Amplifying the curriculum: Designing quality learning opportunities for English Learners.* Teachers College Press.

Planning Science Instruction for Emergent Bilinguals

Introduction

As a former high school science teacher and then beginning science teacher educator, I (Edward Lyon) often felt that I was stuck in the world of "thinking and talking about" social justice—a mantra in my credential program, doctoral program, and current institution—without equipping new science teachers with the models and tools they need to advance social justice in practice. Maybe I wasn't doing enough to help science teachers translate "theory" and "knowledge of research-based teaching practices" into their own science teaching context. Maybe it didn't help that I am a White, monolingual male. Regardless of my own perceived struggles, I became reinspired a few years ago when Dr. Deborah Ball visited Sonoma State to work with some of our preservice teachers and illustrated with clear examples how the teacher has discretion in every teaching moment that can, implicitly or explicitly, disrupt or reaffirm patterns of systemic social inequities. Decisions might be as simple as calling or not calling on a student, or they may be more nuanced, such as revoicing a student's contribution using their own language rather than correcting with standard English standard vocabulary and grammar. Suddenly it clicked. Decisions matter, and the impact (positive or negative) of these decisions accumulates over time. Ball was referring mostly to adaptations teachers can make *while* instructing. But there are also countless decisions made ahead of time while *planning* instruction: intentional decisions about learning goals, questions asked, grouping structures, and creating space for all of students' available resources to be used. Teachers can't make all of these decisions in the moment.

While we should and will talk about students as individuals with unique strengths and challenges in this book, we should also recognize and support particular groups (and the intersection of these groups) that continue to be marginalized in science classrooms, including students of color, students from working-class families, and students whose first or home language is a language other than English. This book, *Planning Science Instruction for Emergent Bilinguals*, is designed to support the teaching of this latter group for the larger vision of advancing social justice. To share approaches and tools for weaving in rich and relevant language support with readers, it was an intentional and easy decision for me to reach out to Kelly Mackura, a science teacher with whom I have copublished, copresented, and cotaught over the last 7 years.

After teaching high school and middle school science for over 20 years, I (Kelly Mackura), as a White, middle-class, native-English-speaking educator, more clearly recognize that my "ways of *doing, knowing, talking, interacting, valuing, reading, writing,* and *representing oneself,*" or *Discourses* (Barton & Tan, 2009, p. 51), are implicitly and explicitly privileged and valued in academic settings. As argued by Barton and Tan (2009), these Discourses are "produced and reproduced in social context and reflect their identity" (p. 51). Regretfully, I have spent most of my career teaching without fully examining how my Discourses are privileged while those of many of my students and families are marginalized, undervalued, or dismissed. I now reflect on different aspects of my teaching. What if excellence and success were measured by seeking out and working to understand richly diverse modalities and lived experiences? What if a full range of human experiences and perspectives were valued, represented, and sought after in our schools, curriculum, and learning experiences? I believe our curriculum, our teaching and learning practices, our assessments, and all aspects of our learning communities need to include and represent our student and family Discourses and lived experiences. What we ask them to do, what we engage them in, and how we assess them should meaningfully connect to their needs, interests, and lives. Curriculum should not be a decision made *for* students and families, but rather a dynamic process that happens *with* students and families throughout each unit. As I continue to grow as an educator and leader, I hope I can continue to use my power and privilege to ensure that what I ask my students to do, what I engage them in, and how I assess them meaningfully connects to their needs, interests, and lives.

Planning Science Instruction for Emergent Bilinguals reflects the broader research consensus to move past the notion of just *eliminating* the linguistic challenges involved with learning content, such as science, and instead provide strategic opportunities and support for students to continuously practice the language and literacy at the heart of science learning. Students' language is not just something in need of improving. All of the Discourses, including language, that students bring into the classroom are resources for science learning. Beyond a *resource*, it is also a student's *right* for science teachers to take full advantage of their lived experiences while utilizing and expanding their linguistic repertoire.

We use particular terms, even in the title, intentionally. First, we will use the term *emergent bilingual* throughout the book instead of *English Learner* or *English Language Learner,* which are still more widely used in educational policy and practice. Instead of defining students by how they are classified in the school system or what they are deemed to be *missing* (i.e., as determined by an English language development test), we focus on assets they bring into the classroom—a language (or languages) other than

English as well as features of the English language that they can draw on to learn science. Thus, their *bilingualism* is emerging.

Naturally, planning *science instruction* is also at the heart of the book. But to make sense of and communicate complex science ideas calls upon students to use language in a range of ways. By *language*, we mean a system of words and symbols, organized in a particular way, used as a tool for meaning-making and for communication. As Lee et al. (2013) argue, we must shift our view of language learning away from acquiring language conventions, rules, and structures and toward the development of "language use for communication and learning" (p. 1). Instead of learning language separately from content, language development occurs *in context*—within specific disciplines and for specific purposes and audiences. And this context is where we start thinking more strategically about language and *literacy*. While literacy can mean many things, we focus on *disciplinary literacy*, or the ways of thinking, the skills, and the tools that are used by experts in the disciplines (Shanahan & Shanahan, 2012). In particular, we focus on how emergent bilinguals make meaning of or through texts. Not only should language be developed in the context of authentic science, but also in a particular student context by leveraging the diverse lived experiences and resources that students bring into the classroom when learning science. And this is where the focus on "rich" and "relevant" enters the scene.

When we think about the integration of science, language, and literacy, then the science that is learned becomes deeper, more interconnected, more abundant. It is the difference between asking someone to "define" a *rainbow* and asking someone to "explain with any language, form of representation, or communication tool what causes the phenomenon of a rainbow." But language and literacy become richer as well. Suddenly, there is something to read, write, or talk *about*.

Science, language, and literacy must also be relevant to the lives of those who have been marginalized in school settings for far too long. Just as language is a resource and a right, students' lives outside of the classroom (e.g., personal interests, home/family/community contexts, histories, culture, and even the local ecology) are a resource and a right to be valued and elevated in status. So we can look toward and work with students and their families to plan instruction that is truly relevant to the students who are being expected to learn science. In summary, science teaching that is responsive to a diverse student context requires that science teachers and those working with science teachers put language, literacy, and students front and center alongside science content when planning science instruction. This may take a modest or significant shift in thinking and in practice to reenvision what it means to learn science and what it means to consider language in the context of science learning.

This book is primarily geared for science teachers and we will use the pronoun "you" throughout to address you, the science teacher. More specifically, this book is intended for science teachers who teach either in a mainstream or sheltered science class. By *mainstream*, we mean general education classrooms that combine students who are not receiving additional language support with students who are still receiving language support. This contrasts with *sheltered* classrooms, where academic content is taught to a entire class of students still receiving language support. We are saying that it is the science teacher's responsibility, along with English language development (ELD) specialists and others, to support emergent bilinguals' language development. To promote teacher learning about weaving in rich and relevant language support, we will illustrate and deconstruct a middle school science unit as a "shared text" (to align with our literacy emphasis) throughout the book. The hope is that you read, discuss, and apply the science unit while working with, for example, science method instructors, ELD and literacy method instructors, mentors, supervisors, instructional coaches, teacher leaders, and/ or administrators. Although we use a middle school example, you can apply the planning process to grades PK–12. We do not take the stance that this unit is perfect, guaranteed to help emergent bilinguals learn, or should be enacted exactly as described. Instead, you can take away a model to adapt to your own curricular and student context. We hope you take away a research-informed planning process to guide how you plan and think about teaching with emergent bilinguals front and center.

For ease, we employ the pronoun "we" to describe planning for the model unit (e.g., "we chose the standard *Interacting via written English*"). But to be clear, Kelly Mackura is the one who developed, implemented, and refined the unit as part of an MA thesis project in collaboration with her preservice teacher candidates and departmental colleagues. This model unit was grounded in research carried out by Edward Lyon with his colleagues as part of the National Science Foundation–funded project Secondary Science Teaching with English Language and Literacy Acquisition (SSTELLA) Project that will be described in Chapter 1. As a science mentor teacher, Mackura collaborated on both the SSTELLA Project and the Accelerating Academic Achievement for English Learners (AAAEL) Project, led by Kelly Estrada and funded by the U.S. Department of Education. These collaborations allowed Mackura to coconstruct planning tools to develop and implement the model unit.

Planning Science Instruction for Emergent Bilinguals is organized around three parts. In Part I we lay out the foundations for how we approach teaching science to emergent bilinguals. Specifically, Chapter 1 provides background on who emergent bilinguals are in the United States, Next Generation Science Standards (NGSS) in relation to teaching emergent bilinguals, and a practice-based approach to teacher learning. Chapter 2 acts as a primer for understanding the theory underlying science, language, and literacy–integrated instruction, including an instructional framework developed and researched in the aforementioned SSTELLA project. We then introduce the model unit, titled Ecosystem Interactions and Resources, in Chapter 3.

Part II deconstructs the model unit and provides guidance for planning instruction at the unit level, such as knowing emergent bilinguals and their families (Chapter 4), unpacking NGSS and curricular resources (Chapter 5), weaving in English language arts and English language development standards (Chapter 6), and anchoring a unit around a phenomenon, text, and assessment (Chapter 7).

Part III then moves into a progression of conceptual understanding and language use. Chapter 8 shows how to analyze the anchor text and further unpack language demands that will support the concepts being learned during a unit. We then illustrate how you can plan a progression of interpretive language (Chapter 9), collaborative language (Chapter 10), and productive language (Chapter 11). These chapters also include sample dialogue and student work from when it was implemented in Mackura's classroom.

Finally, we conclude with recommendations for various types of collaborations to strengthen planning, such as between teachers and with university educators, preservice teachers, students, and families. In the appendices, we also provide unit frames from three additional units at the high school level as well as three more detailed lesson plans from the Ecosystem Interactions and Resources unit. While reading, discussing, and applying, we hope you take the time to reflect on your own stance and practice toward teaching science to emergent bilinguals and always put these students front and center when making planning decisions.

There are many individuals whose ideas and support were instrumental in bringing this book to life. They include the entire SSTELLA Project team—particularly Trish Stoddart, Sara Tolbert, Jorge Solís, Doris Ash, and George Bunch—for the intellectual engagement that led to our SSTELLA Framework. Kelly Estrada, Jackie Guilford, and other AAAEL Project faculty members helped develop and disseminate tools that our teachers can use to analyze and use texts in content area classrooms. Suzanne Garcia, Anthony Bardessono, Kayla Anderson, and Annette Bustamante all contributed science unit frames and learning activities to this book.

Part I

FOUNDATIONS FOR TEACHING SCIENCE TO EMERGENT BILINGUALS

CHAPTER 1

Reenvisioning Science Teaching for Emergent Bilinguals

In this first chapter, we introduce who emergent bilinguals are, the importance of Next Generation Science Standards for emergent bilinguals, and how we approach teacher learning through core teaching practices. In this chapter, consider the following guiding questions:

- In what ways are emergent bilinguals a diverse group of students?
- Why are science practices considered "language-rich" practices?

EMERGENT BILINGUALS IN PK–12 SCIENCE CLASSROOMS

Who is this book specifically designed to impact? While the teaching approaches and tools will be useful for working with all students regardless of language proficiency or status, it is beyond just "good teaching." We are calling attention to teaching approaches and tools that are even more critical for the group of students whom we call emergent bilinguals. This is a diverse group. Some emergent bilinguals are new to U.S. schools and speak virtually no English (i.e., the dominant language of instruction). Some emergent bilinguals are placed in sheltered or mainstream classrooms where they can receive language development support while learning academic content. And then other emergent bilinguals have advanced in their English language proficiency so that they exit a language assistance program (i.e., are redesignated) and may seem to need little support, but yet have not had the full opportunities to do academic work in English or a language other than English.

Emergent bilinguals are also diverse in terms of lived experiences, interests, and academic background (Faltis & Ramírez-Marín, 2015; Kibler et al., 2014). There are approximately five million emergent bilinguals in PK–12 classrooms in the United States—just over 10% of the school-age population. These numbers do not include students who enter a school initially fluent in English (and speak a language at home other than English) or who have exited from language assistance services. This percentage varies by setting and by state. For instance, schools in urban settings typically serve a higher percentage of emergent bilinguals than schools in suburban or rural settings. California consistently has the largest number and proportion of the school-age population who are emergent bilinguals (19.2%), followed by New Mexico, Nevada, and Texas in terms of proportion. However, the most recent growth in the percentage of emergent bilinguals has occurred in many Midwestern states, such as Kansas (from 3.3% in 2000 to 10.3% in 2017). The vast majority of emergent bilinguals were born in the United States.

Unfortunately, if emergent bilinguals are not supported, which includes valuing and utilizing their language and identities, the consequences can be devastating. For over 50 years, emergent bilingual achievement has fallen behind that of native English speakers in science and literacy. In fact, the gap increases from elementary school to secondary school (National Center for Education Statistics, 2016). Besides gaps on standardized tests, emergent bilinguals are less likely to graduate from high school than English-only minoritized Latinx, Black, or low-income students (DePaoli et al., 2017). Emergent bilinguals are also less likely to pursue advanced degrees in science, technology, engineering, and mathematics (STEM) subjects.

This book focuses on and uses examples from secondary (i.e., Grades 6–12) classroom settings. However, the ideas we discuss can and have been applied to elementary school settings. In fact, some of the research base for this book was initiated in an elementary school setting, and was then reframed around the more abstract and complex science ideas, practices, and texts that emergent bilinguals would likely encounter at the secondary level.

Categorizing these diverse students into a single group has advantages and potential consequences. Through classroom-level, district-level, state-level, and national assessments, we can disaggregate students to identify achievement gaps. Identifying learning gaps and general challenges has led to increased funding for research and professional learning to improve the teaching for and learning of emergent bilinguals. However, categorizing can also lead us to falsely assume that emergent bilinguals are a monolithic group. This can be especially problematic when deficit assumptions are ascribed to that group, such as the idea that emergent bilinguals are not capable of learning rigorous science or participating in class discussions due to their limited English proficiency.

Research has consistently shown that emergent bilinguals can learn academic content, such as science, without being fully proficient in English. Additionally, what is "relevant" to emergent bilinguals might lead to generic views of culture ascribed to the race and/or ethnicity to which students identify. For instance, instead of just assuming that every emergent bilingual whose family speaks Thai enjoys and just routinely eats pad thai, teachers can learn from their students about the rich history of Thai cuisine along with evolving more modern Thai foods. Teachers can also learn about non-Thai food that these students and their families might enjoy! Research has shown how, without support, cultural awareness, and tools from the beginning, English-only speaking science teachers rarely go beyond these static and often stereotyped views of culture when planning relevant instruction for emergent bilinguals (Tolbert & Knox, 2016).

Science teachers consistently report being underprepared to support emergent bilingual students (Banilower et al., 2018; Faltis & Valdés, 2016; Salloum et al., 2020). This finding is understandable. Science teachers might have originally learned models of teaching science to emergent bilinguals that more recent research has critiqued and revised. Science teachers might implicitly or explicitly hold onto the stance that they are teachers of content, despite the fact that scientific argumentation, reasoning, and communication require a multitude of specialized written and oral literacy practices. You might be able to engage in the language and literacy related to science practices, but not know how to make instruction accessible for emergent bilinguals or promote students' language and literacy development. Fortunately, new research and priorities have been developed that can help educators rethink (1) who emergent bilinguals are rethink, (2) what emergent bilinguals are able to learn and do in science, and (3) learn or relearn knowledge and tools around language acquisition and how students use language to learn science (National Academies of Sciences, Engineering, and Medicine, 2018). Chapter 2 provides a primer on the theoretical foundations and research base for this synergistic relationship between language and science.

NEXT GENERATION SCIENCE STANDARDS AND WHY THEY MATTER FOR EMERGENT BILINGUALS

What are we proposing that emergent bilinguals learn and do in PK–12 science classrooms? We use the principles established in *A Framework for K–12 Science Education: Practices, Crosscutting Concepts, and Core Ideas* (National Research Council [NRC], 2012) and the resulting *Next Generation Science Standards: For States, By States* (NGSS Lead States, 2013a) to guide what science all students in the United States should be learning. A majority of states have adopted the Next Generation Science Standards (NGSS) or have adopted state standards based upon recommendations from *A Framework for K–12 Science Education*. It is beyond the scope of this book to examine all the underlying research and rationales behind these documents. Rather, we address a few noteworthy principles that are most pertinent to the teaching of emergent bilinguals.

The ideas and language of NGSS, such as three-dimensional learning, phenomena, storylines, science practices, engineering practices, and crosscutting concepts, look remarkably different from previous state standards and even national documents such as the *National Science Education Standards* (National Research Council, 1996). Like all standards, NGSS is not a stand-alone curriculum, but rather a clearly delineated list of learning (or performance) expectations that a community of science educators, science education researchers, scientists, engineers, and other stakeholders has agreed that students should meet throughout their K–12 science education. NGSS Appendixes expand on its content and organization and provide some guidance for how NGSS can be implemented to teach a diverse group of students (NGSS Lead States, 2013b). NGSS still has its shortcomings and limitations, including lack of more explicit attention to equity and multicultural science education (Rodriguez, 2015). But it marks important progress from earlier reforms by drawing on more recent evidence-based research about science teaching and learning. These views, which are compared to more "traditional" thinking in Figure 1.1, matter greatly for how we think about planning science instruction that weaves in rich and relevant language for emergent bilinguals.

Science as Practice

NGSS performance expectations intertwine core science ideas (e.g., interdependent relationships in ecosystems), science and engineering practices (e.g., arguing with evidence about a core idea), and crosscutting concepts (e.g., cause and effect) instead of separating scientific knowledge (e.g., knowing what a food web is) and scientific process skills (e.g., knowing

Figure 1.1. Comparing the Traditional View of Science Teaching/Learning to Views Laid Out in *A Framework for K-12 Science Education* (NRC, 2012)

Traditional View of Science Teaching/Learning	The Framework's View of Science Teaching/Learning
Students go through developmental stages and can only reason/think scientifically when in secondary school.	Children are born investigators.
Focus is on discrete pieces of scientific knowledge.	Focus is on core ideas in science.
Focus is on scientific process skills ("the scientific method").	Focus is on scientific and engineering practices.
There is a disconnect between content and process.	Science and engineering require both knowledge and practice.
There is a limited empirical basis to curricular sequence.	Understanding develops over time.
Science replaces (or avoids) students' interests and experiences.	Science connects to students' interests and experiences.
Promotes either equity *or* excellence.	Promotes equity through excellence (i.e., rigor).

how to observe carefully). In other words, science is seen as a practice. As described by Stroupe (2015), viewing science as practice "involves a reframing of teaching and learning expectations away from the memorization of information to progressive engagement in authentic disciplinary work." The words *"engagement"* and *"authentic"* are key. It is not enough for students to know "about" a science practice such as planning and designing experiments; it is necessary for them to become an active participant in the practice whereby students use conceptual understanding, language (e.g., forms of communication, vocabulary, symbols), norms, and tools in ways authentic to the scientific community. On the surface, it may seem challenging and even unfair for emergent bilinguals, who are still mastering English and hopefully also expanding on their first language, to successfully engage in such abstract and linguistically demanding activities. Yet we are arguing the opposite. When supported, science practices become the prime context for developing language and literacy because of how language is put in a context and used for a purpose.

Connecting to Student Interests and Experiences

NGSS takes the view that science instruction should connect to students' interests and experiences, rather than avoiding or replacing them. But NGSS alone does not provide the guidance to make meaningful and relevant connections, nor does an educator's own interests about science. In Chapter 4 we will lay out strategies for you to better understand your own students' home, family, and community context in order to ensure that everything science teachers do, even when unpacking standards, is informed by who students are—so that science learning is contextualized to students. This is important for building on the linguistic resources of students as well. In later chapters, we will show you how to strategically plan opportunities for students to communicate science ideas in a variety of tasks and for multiple audiences and purposes, which leverages emergent bilinguals' linguistic repertoire. For example, an emergent bilingual might already have experience with building websites, which requires awareness of how users would access and navigate the content. Thus you could plan an assessment where students have the option of designing a website to defend an argument about reintroducing certain species into a population—as opposed to just asking a single prompt to defend an argument in paragraph form.

Equity Through Academic Rigor

Lastly, the traditional view (column 1 of Figure 1.1) assumes that for emergent bilinguals to *succeed* in science (e.g., learn what they are supposed to learn), that then means diminishing academic rigor. This is an extremely problematic assumption prone to perpetuating patterns of inequity, since they would be denied opportunities to learn and advance in their coursework. It can amount to tracking where emergent bilinguals remain in sheltered classes and never get the opportunities or support to benefit from mainstream and honors or AP science classes.

Rather, NGSS and this book take the stance that equity happens through excellence. Emergent bilinguals need appropriate opportunities and support to engage in rigorous science to expose them to the varied uses of language called for in science and engineering practices and deeper core science ideas. By recognizing and using NGSS when planning science instruction, you are already on the path to disrupt patterns of inequity. Rigorous expectations for students combined with strategic support to engage in rich thinking and language use will accelerate science learning as well as language and literacy development. NGSS, along with student context, provides the frame in which to design these supports.

Science Practices as Language-Rich Practices

Each of the eight NGSS science and engineering practices involves an array of tasks that students would learn and do to engage in the practice successfully. In many ways, these tasks are what define the shift toward science as practice—students are not just learning to recite back knowledge, but rather learning to apply and communicate core ideas in ways that are authentic to *doing* science and engineering. Lee et al. (2013) outlined the different ways in which students would engage in each science and engineering practice cognitively and linguistically. For example, to "construct scientific explanations," students engage in the following internal sensemaking (or cognitive) activities:

- *Developing* explanations
- *Analyzing* the match between explanation or model and a phenomenon or system
- *Revising* explanations based on input of others or further observations
- *Analyzing* how well a model and evidence are aligned

Other activities associated with "scientific explanations" include:

- *Comprehending* questions and critiques
- *Comprehending* explanations offered by others
- *Communicating (orally and in writing)* ideas, concepts, and information related to an explanation of a phenomenon or system (natural or designed)
- *Providing* information needed by listeners or reader
- *Responding* to questions by amplifying explanation

- *Responding* to critiques by countering with further explanation or by accepting as needing further thought
- *Critiquing or supporting* explanations offered by others

There is something different about the second list. While the first involves cognitive (or internal connecting, reasoning, and representing ideas) tasks to develop, analyze, and revise scientific explanations, the second list involves acts of receptive (e.g., comprehend) or productive (e.g., provide information, respond) modes of communication through language. Both cognitive and linguistic tasks are necessary, reflecting the reciprocal and synergistic relationship between learning science and developing language and literacy. Students need to comprehend, communicate with others about, and structure explanations to engage with and learn how to construct explanations, which allows them to make sense of a core idea. Simultaneously, the act of making sense of an explanation provides the opportunity for developing language through contexts that are authentic to science, instead of developing language divorced from contexts. In essence, science and engineering practices *are* language-rich practices. Understanding this richness in science and language, you can start developing more awareness of the language that will be involved in specific performance expectations and allow them to create more intentional support for students, especially emergent bilinguals.

Opportunities for Literacy Through NGSS

It is not an easy task for you to implement new science standards, especially standards that depart from the highly prescriptive and discrete learning goals of the past. Added to that challenge, you are increasingly expected to support students' reading and writing of a broad array of print and digital texts as they are learning science. But supporting students' literacy, or even *bi*literacy, in the context of science better positions emergent bilinguals to argue with evidence, analyze data, develop models, and communicate information—not necessarily because of students' abilities to make sense of complex science ideas, but because of the language structures and purposes (beyond grammar and vocabulary) that are central to carrying out these practices.

Herein lies an opportunity. NGSS was developed just after the National Governors Association and the Council of Chief State School Officers released Common Core State Standards in Math and English Language Arts in 2010 to guide "college and career readiness" for all learners. Common Core also marked a notable curricular shift, such as the emphasis on critical thinking and literacy in the context of academic disciplines such as science. Science and engineering practices actually provide a natural vehicle with which to *promote* literacy as called for in Common Core. This appears to be a timely and reciprocal relationship. Science teaching benefits from guidance around supporting language and literacy found in Common Core ELA and in English language development standards that will be discussed later. Supporting literacy benefits from standards that emphasize literacy in a disciplinary context for particular purposes, such as science practices. The approaches used in this book and the theory guiding them really come down to this synergistic and reciprocal relationship between learning science and developing language and literacy.

SCIENCE TEACHER LEARNING THROUGH
CORE TEACHING PRACTICES

Finally, how do we approach *science teacher* learning? Just as we can identify the key intellectual activities of scientists (i.e., science practices), we can identify key practices that teachers can use to help diverse students. Education scholars have collaborated across disciplines to identify a set of "core" teaching practices that even novice teachers can use to guide learning (Grossman et al., 2009; McDonald et al., 2013; Windschitl, Thompson, et al., 2012). In fact, the emphasis on core practices ensures that science teacher education programs support teachers in "actually taking on the role of the teacher . . . while receiving feedback on their early efforts to enact a practice" instead of just "discussing what one might do as a teacher" (Grossman et al., 2009). While most teaching practices focus on what happens during instruction (e.g., eliciting student ideas), this book intentionally focuses on *planning* rather than just what to do during instruction. Strategic planning is what creates the spaces and opportunities for effective instruction, assessment, and student learning.

Tools are an important aspect of a practice-based approach. Davis and Krajcik (2005) refer to materials, including lessons, assessments, and student work, used to promote teacher learning as educative curriculum materials. Starting in Chapter 4, we present a series of Planning Tools with which science teachers can engage in targeted planning practices such as weaving in Common Core ELA and ELD standards. By focusing on planning practices, we hope to support teachers in making the complexities of integrating science, language, and literacy during instruction more fluid and automatic. We also emphasize how the planning practices can be learned and rehearsed. But to better deconstruct and approximate those practices that may lead to improved science learning and language and literacy development for emergent bilinguals, it helps to understand the theoretical and empirical research forming the foundation for this book. In the next chapter, we highlight key findings and insights from this research and how it will guide our planning.

CONCLUDING REMINDERS

- The group that we refer to as emergent bilinguals is culturally, linguistically, socioeconomically, geographically, and academically diverse.
- Next Generation Science Standards represent shifts in what we expect all students to learn and do in science—shifts that can provide emergent bilinguals with access to rich and relevant science instruction.
- Science and engineering practices are language-rich practices that can further support emergent bilinguals' language and literacy development.
- The science unit and associated planning tools we will use throughout the book can help guide in weaving in rich and relevant language support for your own science units.

CHAPTER 2

A Research Primer for Integrating Science Learning With Language and Literacy Development for Emergent Bilinguals

In this chapter we describe the theory and research grounding our approach to the planning of science instruction that weaves in rich and relevant language support for emergent bilinguals. In particular, we emphasize the central role of language, coupled with the social, cultural, and historical context, in teaching and learning science. We also describe how to move away from traditional notions of academic language by focusing on how students use language to communicate discipline-specific ideas (referred to as "language demands"). We then make connections to the Secondary Science Teaching with English Language and Literacy Acquisition (SSTELLA) instructional framework as a way to begin bridging theory to practice. This may be a useful chapter both before reading the subsequent practice-based chapters and after reading them—to reflect and deepen your insight about why you should plan through the suggested approach. During the chapter, consider the following guiding questions:

- How do disciplinary and sociocultural perspectives inform the teaching of science to emergent bilinguals?
- How can we rethink the notion of "academic language" in ways that better support emergent bilinguals?
- What is the role of language demands in helping to plan science instruction for emergent bilinguals?
- What is the role of the SSTELLA Framework in helping to plan science instruction for emergent bilinguals?

THEORETICAL FOUNDATIONS FOR INTEGRATING SCIENCE
WITH LANGUAGE AND LITERACY DEVELOPMENT
FOR EMERGENT BILINGUALS

Back in the 1980s Cummins (1980) proposed that teachers of emergent bilingual students should recognize two distinct language domains—Basic Interpersonal Communications Skills (BICS) and Cognitive Academic Language Proficiency (CALP). Whereas BICS referred to social language (imagined as language used on the playground or in informal conversations), CALP referred to the language to be used to do academic work—or academic language. He proposed that a role of teachers is to support emergent bilinguals in developing CALP in disciplinary-specific contexts. At the same time, Krashen (1981) argued that students acquire a second language (e.g., English) through meaningful and supportive communicative interactions in the second language. The idea behind "communicative interactions" is to provide instruction where the meaning could be understood without necessarily understanding every word uttered. The focus on this comprehensible language is a key aspect of the Sheltered Instruction Observation Protocol (SIOP) (Echevarria et al., 2008). SIOP still remains a commonly used approach in teaching science to emergent bilinguals in both *sheltered* (exclusively emergent bilinguals) and *mainstream* (combination of emergent bilinguals and nonemergent bilinguals) settings. SIOP includes strategies such as building on students' prior knowledge and experiences, front loading vocabulary, using visual representations, modifying texts, and directly translating English into students' native languages to make content comprehensible for emergent bilinguals (Goldenberg, 2013).

In light of new research and reforms, scholars have critiqued the SIOP model for its behaviorist underpinnings and disconnect from discipline-specific learning (Crawford & Adelman Reyes, 2015). More recent teacher education research for teaching content to emergent bilinguals has departed from exclusively sheltered strategies and instead invoked disciplinary and sociocultural perspectives. A disciplinary perspective argues that supporting emergent bilinguals' language and literacy is central to content-area learning, not an "add-on" to it (Bunch, 2013). For example, instead of separating language learning goals from content learning goals, as is common in the SIOP approach, goals should *integrate* language and content. Separate goals might be stated as "write and defend a claim with evidence" [language] and "demonstrate how genetic variations may result from new genetic combinations through meiosis" [science content goal]. An integrated goal would look like the following: "Make and defend a claim based on evidence that inheritable genetic variations may result from new genetic combinations through meiosis." This integrated goal is part of an actual Next Generation Science Standard performance expectation. Integrated goals call upon students to demonstrate science through specific uses of language (in this case arguing with evidence).

A disciplinary perspective also recognizes that how language is learned and used differs across disciplines. Mathematicians, scientists, and social scientists all read and interpret texts to do math, science, or history. A text is anything that conveys meaning to the reader: a brochure, website, Instagram, video, graph, article. Yet the kinds of texts experts read and how they interpret and make meaning from them differs across the disciplines (Shanahan & Shanahan, 2012). Thus, instead of generic language goals that can be taught the same way regardless of discipline, you can support emergent bilinguals by helping them understand how to use language, such as arguing with evidence, in the context of the specific discipline, such as biology.

A sociocultural perspective pushes further and recognizes that learning cannot be fully explained without understanding the social, cultural, and historical contexts. For one, this means considering how science is taught in our schools and how this may be in conflict with the ways of being of our students from nondominant backgrounds. As Barton and Tan (2009) argue, "valuing diverse funds of knowledge and Discourse as legitimate science classroom resources positions minority students as rightful experts of certain knowledge directly related and applicable to school science" (p. 52). In Chapter 4 we will highlight tools for gathering and integrating students' out-of-school lived experiences and their ways of communicating, representing, and being.

Together, disciplinary and sociocultural perspectives provide the theoretical grounding for reenvisioning what it means to consider and support language and literacy in content-area instruction. A strong empirical research base demonstrates that integrating science instruction with targeted language and literacy development improves content learning and literacy for emergent bilinguals (August et al., 2014, Lara-Alecio et al., 2012; Llosa et al., 2016). One explanation for this promising research is that when taught authentically, content-area instruction provides a natural vehicle in which to integrate language and literacy. By "authentic," we mean that students are *doing* science in the classroom. They are actively using conceptual ideas, language, norms, and tools in ways authentic to the scientific community. But that calls on you to identify the specific ways students would be using language. Next, we describe the notion of language demands, which clarifies how "academic language" has more to do with the varied ways to communicate science ideas than just precise, technical science vocabulary and grammar.

LANGUAGE DEMANDS AND RETHINKING THE NOTION OF "ACADEMIC LANGUAGE"

Reenvisioning the role of language and literacy in learning science has implications for instructional planning. In the aforementioned SIOP model, language supports, such as visuals, graphic organizers, and sentence frames, are commonplace to make content comprehensible. Those same strategies are useful, but you should use them in support of the larger goal to provide emergent bilinguals with widespread opportunities to develop language and literacy as they make sense of complex science ideas. By widespread, we mean different modes (writing, reading, speaking, listening, drawing, gesturing), student interactions (pair-share, small group, whole class, via digital tools), texts (articles, graphs, simulations, lab reports), and languages. Emergent bilinguals also need targeted support to understand relationships and differences among various language forms and registers (e.g., discussing findings within a collaborative group vs. writing a scientific report) (Hakuta et al., 2013; Lee et al., 2013). You can draw on common English language development frameworks and standards to guide this targeted support. We will show how to do this in later chapters.

Language demands are the varied functions and features of language that pertain to making sense of and communicating discipline-specific ideas. The purpose of this term is to depart from the dichotomous view of language as either exclusively "social" or "academic." Instead, all students, but especially emergent bilinguals, can use language that is more familiar in their everyday lives just as well, and in many cases more powerfully, to make sense of and communicate ideas than the language that they will be more likely to encounter in science texts. Instead of thinking whether words and texts sound "smart" or not, think what purpose are words and texts serving to learn and do science.

The term *language demand* still might be misleading. It does not necessarily mean that the language is *hard* for students, just that students use language for some *purpose* (thus it does place a cognitive demand on them). Figure 2.1 describes and provides examples of four categories of language demands that will be referred to throughout the book: language function, discourse, syntax, and vocabulary. These categories become useful when unpacking standards (Next Generation Science Standards, Common Core, English language development) to understand how students would be given opportunities and support to communicate discipline-specific ideas.

It helps to begin with the overarching purpose of using language, in other words, a language function. For example, you might have the goal for students to analyze data around the relationship between pressure and volume of a gas. The language function in this case is "analyze." By starting with the language function, you can then think about how students analyze by using technical and nontechnical words, phrases, and symbols (i.e., vocabulary such as the symbols P, V, n, R, and T), which are organized in particular ways (i.e., syntax such as the ideal gas law equation $PV=nRT$), and then used to talk, write, represent, or reason to convey meaning (i.e., discourse such as representing relationships on a graph or writing the findings in a lab report). Texts produced by students can be written (lab report), graphical (chart or graph), or digital (video, simulation). Texts become a tool for communicating specific disciplinary purposes, such as analyzing data about the changing conditions of gas samples.

Figure 2.1. Categories of Language Demands

Language function: Active use of language for a particular purpose and audience (i.e., what you *do* with language)

Examples

- *Compare* photosynthesis to cellular respiration
- *Explain* what causes genetic variation
- *Analyze* data around relationship between pressure and volume of a gas

Language Features		
Discourse: How scientists or engineers talk, write, and reason about, for a particular purpose and audience (i.e., tools that convey meaning)	**Syntax:** How we organize words or symbols in science to convey meaning (i.e., tools that organize)	**Vocabulary:** Technical and nontechnical words, phrases, and symbols
Examples	**Examples**	**Examples**
• *Coordination of claim–evidence–reasoning in an explanation* • *Format of a lab report* • *Representing data in a graph*	• *Use of symbols to indicate balanced equations* • *Conditional statements to make predictions: if heterozygous yellow peas are crossed with homozygous recessive green peas, what resultant offspring genotypes could occur?*	• *Chloroplast* • *Force* • *Reactant* • *PV=nRT*

To preview the type of language demands called for in Next Generation Science Standards, Common Core English Language Arts Standards, and English language development standards, we have displayed in Figure 2.2 a 7th-grade student writing sample from the unit that will be discussed throughout the book. For one, the primary *function* of this writing was to *explain* (or provide a causal account), with evidence, for changes in sea urchins, otters, and kelp in a particular kelp forest over time. To accomplish this function, the student needed to know how to convey meaning in ways that are authentic to science and the scientific community. During the unit, he learned about and demonstrated the claim–evidence–reasoning (CER) structure in his explanation. He also had support in writing and organizing sentences and visuals that help convey this meaning. The writing starts with a claim: "In 1970, Sea Urchin populations decreased." Through previously used sentence frames, he was supported in how to describe reasoning—"because sea otters consume sea urchins"—which applies what he had learned about predator–prey interactions in this ecosystem. He was also able to show this relationship visually and reference the graph that included the data to support the claim. The structure of sentences and visuals and use of transitions and phrases to denote causation (e.g., "because") are examples

Figure 2.2. Sample Emergent Bilingual Explanation

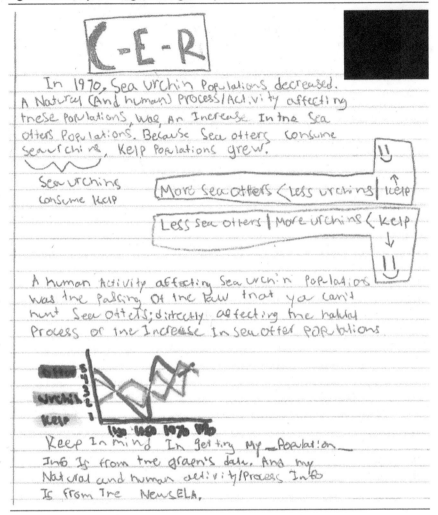

of discipline-specific syntax. Finally, he was able to appropriately use vocabulary that was targeted during the unit (e.g., natural process, consume).

All students will come into the classroom at a different reading and writing level. And these levels vary by what they are reading and writing—a scientific explanation is very different from writing a personal narrative. The key is to maintain rich expectations for language and literacy and consider the level and type of support provided depending on what the student can do with language and what linguistic resources they bring to the classroom.

TRANSLANGUAGING AND THE FLUIDITY OF LANGUAGE USE

While we might be eventually expecting more formal discourse, syntax, and vocabulary from students, we are promoting planning that provides a space for a wide range of language use to serve as entry points and bridges so that emergent bilinguals can gain access to these dominant forms of language while sustaining their own linguistic identity. We value the diversity of ways that emergent bilinguals might, for example, construct an argument. The diversity of their ways of communicating (or linguistic repertoire) provides the resources for meaning-making, but also connects to who they are—their identity in and out of the classroom. When recognized, valued, and supported, you can even expand emergent bilinguals' linguistic repertoire. Imagine an emergent bilingual having the flexibility to explain a scientific model in English, in Spanish, or a combination of both languages. Essentially, they would be equipped with a greater linguistic toolkit in which to make meaning and communicate.

Buxton and Caswell (2020) emphasize the *fluidity* of language and literacy in learning science. Students use language to communicate and make sense of science ideas differently depending on the topic (what), audience (who), and mode of communication (how)—collectively referred to as the "communicative context," or "register." In a science classroom, students are constantly called upon to shift registers. Consider the varied language choices and uses needed for students to actively participate in a small-group discussion where they are modeling meiosis on a whiteboard or with pipe cleaners. The social and more colloquial forms of language are more prevalent and perhaps even more useful for exchanging ideas than the more technical and discipline-specific language that is often ultimately expected by teachers. Students might automatically shift back and forth from formal to informal (and everything in between). For some emergent bilinguals, this might include strategic shifts in the language itself (e.g., English to Spanish). García (2009) and other scholars refer to this practice where students shift fluidly across linguistic registers as *translanguaging*.

You can support translanguaging in several key ways. First, you can provide norms, expectations, and affirmations that encourage students to talk and make meaning with one another using whatever language helps. In other words, in the collaborative meaning-making process, emergent bilinguals should not feel obligated to use technical words and phrases that could inadvertently hinder the sensemaking process. In one videotaped classroom that we frequently show to preservice science teacher candidates, a high school student is asked to share his response to the prompt students had been practicing around using "before" and "after" phrases to describe a sequence of geologic events. When prompted, the student asks for clarification, "Do I say it the *smart* way or *my* way?" This one sentence says a lot about the students' perceptions about what it means to talk and learn in a science classroom. It also signals an important moment where you might inadvertently reinforce this student's perception and devalue some of the linguistic resources he has for meaning-making. Instead, you might encourage all of the ways that he can participate and value his way as legitimate and smart in its own right.

When possible, you can provide texts, video subtitles, and directions in English and languages other than English. You can also encourage emergent bilinguals to respond orally or in writing in any language. Even if you are monolingual, there are other students, other teachers, English language development coaches, and Google Translate to assist you. Widespread student interaction provides more opportunities and access points for students to make sense of science ideas and deepen them (Hawkins, 2004; Michaels & O'Connor, 2012). In addition to oral language during collaboration, gestures (imagine students using their hands to represent chromosomes being separated from each other) and manipulatives (e.g., pipe cleaners) become useful tools for emergent bilinguals to make meaning with others. You might then call upon students to write up investigation findings through a lab report using a very prescriptive format for arguing with evidence and style that differs considerably from the language used in small-group discussions. This constant shifting is a challenge, but when supported, it is an opportunity to help emergent bilinguals expand their linguistic repertoire. If you intentionally plan norms, structures, and spaces where students can move across registers and languages, then students may be more likely to see that their way *is* a smart way as long as it is helping them make and communicate meaning.

THE SECONDARY SCIENCE TEACHING WITH ENGLISH LANGUAGE AND LITERACY ACQUISITION (SSTELLA) FRAMEWORK

Edward Lyon and his colleagues collaborated with university science teacher educators in the National Science Foundation–funded Secondary Science Teaching with English Language and Literacy Acquisition (SSTELLA) Project to develop a cohesive instructional framework that can better describe the kinds of science teaching practices that are science-, language-, and literacy rich and relevant (see Lyon, Stoddart, et al., 2018). As described throughout this chapter, instead of the predominant sheltered approaches for teaching content to emergent bilinguals that emphasize the linguistic features that make science learning challenging for them, SSTELLA emphasized a disciplinary and sociocultural approach. In this disciplinary and sociocultural approach, you can leverage science practices and discourse, as well as knowledge of students' out-of-school and home/family experiences and assets, to support emergent bilinguals in both learning science and developing language and literacy.

SSTELLA's primary goal was to strengthen how grades 6–12 science teachers were prepared to teach emergent bilinguals through a restructured university teacher preparation science methods course. But we also collaborated with inservice mentor science teachers with whom the preservice teachers were placed. Through this collaboration with mentors, including Kelly Mackura, we were able to develop exemplary videotaped learning activities showcasing key teaching practices in action. The resulting SSTELLA instructional framework consists of four interrelated dimensions, described next, which align with Next Generation Science Standards, Common Core State Standards (CCSS), as well as commonly used English language development Standards.

Position All Students as Sensemakers

By "sensemakers," we refer to a perspective posited by Warren et al. (2001) working in racially, linguistically, and socioeconomically diverse classrooms in which students used "a varied complex of resources, including practices of argumentation and embodied imagining, the generative power of everyday experience, and the role of informal language" (p. 532) in making sense of complex science ideas. This means that instead of replacing everyday language, experiences and thinking with "scientifically correct" language and thinking, you can legitimize the varied forms of language and thinking, especially from nondominant and marginalized student groups, including emergent bilinguals (Ash, 2004).

Be careful not to fall into what Lemmi et al. (2019) call a language-exclusion ideology, where you only value technical science vocabulary, proper grammar, and other perceived "high-status" forms of language. Instead, we hope you take a language-inclusion ideology and value multiple forms of language (e.g., home language, dialects, colloquial discourse) as resources for science and language learning.

Science-, language-, and literacy-rich and relevant instructional planning draws on these dimensions first and foremost by framing a unit of instruction around an anchor phenomenon (i.e., observable event or real-world problem/application). From this framing, you can plan learning activities that engage all students in science and engineering practices, such as through developing, using, and explaining scientific models related to the anchor phenomenon. While planning these activities, you would put into place structure and support that not only make content more accessible (e.g., using a graphic organizer to develop a scientific model) but also leverage student and family experiences and other assets for sensemaking.

Facilitate Productive Disciplinary Discourse

We have already described scientific discourse as a category of language demands—how scientists or engineers talk, write, and reason about, for a particular purpose and audience (i.e., tools that convey meaning). We have also alluded to the importance of talk that is disciplinary in nature and supports sensemaking and problem solving—thereby productive. To create these spaces, you can strategically plan talk moves, norms, structures, and resources (e.g., time to think or write before talking, strategies to ensure participation from all students, strategic grouping) to support emergent bilinguals in exchanging and elaborating on information/ideas with each other (in partners, small groups, or with the whole class), resulting in sustained discussion. Fostering productive disciplinary discourse elevates the rigor of the learning experience for all students, including emergent bilinguals, when their ideas, questions, and reasoning are publicized and refined (Thompson et al., 2016).

Scaffold Language and Disciplinary Literacy

We have also discussed the critical role science classrooms play for developing emergent bilinguals' language and disciplinary literacy. To carry out this role in practice, you can use *scaffolding*—plan carefully sequenced goals, learning activities, and assessment activities that support emergent bilinguals in ways that move them toward language and literacy development (Walqui & Bunch, 2019). For instance, you can plan learning activities in which emergent bilinguals use varied modes (e.g., developing a visual model, oral presentation, discussing in a small group) to represent thinking about a scientific phenomenon being explored. In these activities, you can use strategies to support development and use of vocabulary, symbols, and discourse to communicate these modes.

You can also plan learning activities in which students are supported in reading, writing, and discussing a text. These activities can include attention to the structure, function, and audience of text and how to use linguistic resources (English and/or native language vocabulary, discourse, syntax) to write for a particular purpose and audience. Collectively, scaffolded disciplinary literacy elevates academic rigor as emergent bilinguals contribute their own funds of knowledge and shift across and/or use multiple registers, texts, modes, and languages (Buxton & Caswell, 2020; Kavanaugh & Rainey, 2017).

Contextualize Instruction

From a sociocultural perspective, science learning for emergent bilinguals is enhanced when you make instruction meaningful and relevant, or contextualized (Lyon, Tolbert, et al., 2016). A meaningful and motivational context is important for all learners, and the ability to apply math/science knowledge and skills to real-world problems is an important goal of Next Generation Science Standards. However, everyday events outside of school (current events, community issues, youth activities) are essential to leverage the intellectual, linguistic, and cultural resources emergent bilinguals bring to learn content and develop language/literacy (González et al., 2005).

Contextualizing instruction can be understood as both planning instructional goals and activities around relevant contexts as well as planning the spaces in which emergent bilinguals can contribute their lived experiences and funds of knowledge during instruction (Lyon, Tolbert, et al., 2016). Who the students are, and the context of their family, home, and community, should inform all planning decisions. This is more than just a familiar context, and you must be careful not to make assumptions and use generic notions of culture, or use contexts that are just interesting to the science teachers. Before and throughout the unit, you can plan spaces in which students contribute personal, linguistic, cultural, or community assets that can then be leveraged to help emergent bilinguals develop and use the language demands (e.g., vocabulary, symbols, syntax, discourse) necessary to reason, explain, or argue more precisely.

The SSTELLA Framework can benefit you in planning science instruction for emergent bilinguals, not because it is a "magic bullet" or the singular best approach for planning a science-, language-, and literacy-rich and relevant unit, but rather because the SSTELLA Framework does provide a coherent and research-based rationale that you can apply to your own contexts and priorities. The interplay of SSTELLA themes, modes of communication, and language demands will reappear over and over as we help you weave rich and relevant language support into your science planning for emergent bilinguals.

CONCLUDING REMINDERS

- Sociocultural and disciplinary perspectives argue that supporting emergent bilinguals' language and literacy is central to content-area learning, not an "add-on" to it.
- You can support emergent bilinguals in three particular modes of communication: Interpretive (making meaning of texts), collaborative (communicating *with* others to make meaning), and productive (communicating meaning through texts).
- Language demands (i.e., language functions, vocabulary, syntax, and discourse) help us focus on using language for meaning-making and communication.
- The SSTELLA Frameworks can help you keep coming back to the following central themes when planning: Position emergent bilinguals as sensemakers, facilitate productive disciplinary discourse, scaffold language and disciplinary literacy, and contextualize instruction.

CHAPTER 3

Ecosystem Interactions and Resources

A Sample Unit to Illustrate the Planning Process

In this chapter we will introduce the Ecosystem Interactions and Resources unit. Our purpose is to use this unit to illustrate the planning process. It serves as a shared text for you to read, discuss with others, and apply to your own teaching or work with teachers. We start by describing the students and curricular context—who the students are and the course they are taking. We then summarize the Ecosystem Interactions and Resources unit frame—key features that drive how we sequenced objectives, learning activities, and assessment so that emergent bilinguals progress as they learn science and use language. Finally, we share an outline of the objectives, learning activities, and assessment from the unit so that you have a sense of where the planning process ended. We also introduce you to the set of seven planning tools to guide the planning process. During the chapter, consider the following guiding questions:

- How can you best use the Ecosystem Interactions and Resources unit frame for your own professional growth?
- What is the role of the unit frame in planning science instruction for emergent bilinguals?

STUDENT AND CURRICULAR CONTEXT

Ecosystem Interactions and Resources was planned for and taught in a 7th-grade integrated science class in Santa Rosa, California. Santa Rosa is a suburban setting with approximately 160,000 residents. However, Santa Rosa is surrounded by the overall rural Sonoma County, which features a diverse ecological landscape—from the Pacific Ocean to redwoods to wine vineyards to woodland meadows. The landscape and people have also been dramatically affected by intense wildfires and flooding in recent years. In the particular class that we will discuss, 31% of the students were emergent bilinguals and another 40% were redesignated as fluent in English. The primary language of 69% of our students was Spanish, while two other students had a primary language other than Spanish or English. Half of the students with English as a primary language received dual immersion Spanish/English education during their elementary school years at Cesar Chavez Language Academy. We point this out because of how the student context matters when planning. What is relevant to students in this school differs from what is relevant in other communities, although naturally there are larger issues and contexts (e.g., a global pandemic) that affects us all. The languages and students' proficiency in those languages also matter. In Chapter 4, we will share ideas for gathering an even richer source of knowledge about your students and families.

The course wove together the physical sciences, life sciences, and earth and space sciences in each unit. This integrated course model drew heavily on the 2016 California Science Framework as a curricular resource, which proposed NGSS-aligned instructional modules that wove ideas from the physical sciences, life sciences, and earth and space sciences into a cohesive storyline. The course also uses TCI's *Bring Science Alive!* online resources (see https://www.teachtci.com/science/).

UNIT-LEVEL PLANNING

Generally, we conceive of a *unit* as a cohesive sequence of activities framed around a central conceptual focus. In Figure 3.1 we depict this overarching framing for a 3-week-long Ecosystem Interactions and Resources unit. In the unit frame we identify essential student assets (i.e., what lived experiences,

Figure 3.1. *Ecosystem Interactions and Resources* Unit Frame

Essential Student and Family Assets
- Lived Experiences
  » Mountain biking, playing basketball, playing soccer, skateboarding (visiting skate parks), walking and playing with their dogs, hiking, visiting and playing at local parks near their homes or their own gardens and yard.
  » Students enjoy exploring and discovering new plants and animals in large outdoor spaces.
  » Many students shared visiting the beach as one of their favorite outdoor spaces. At the beach, they enjoy playing in the sand, playing in the water, spending time with family and friends, experiencing sunsets, and feeling happy and calm.

(continued)

Figure 3.1. (*continued*)

- **Linguistic Repertoire**
  - » Students and families enjoy oral interviews and spoken communication as a means of sharing thinking and ideas.
  - » Spanish is spoken at home and in the classroom
- **Home/Family/Community Funds of Knowledge**
  - » Water and trees are highly valued ecosystem resources.
  - » Spending quality time together with family and community is highly valued and often connected to experiences outside in nature, especially when visiting the ocean.

Big Science Idea

Natural processes and human activity affect Earth's resources and ecosystem.

Essential NGSS Standards	Targeted CCSS ELA and ELD Standards
MS-LS2-1. Analyze and interpret data to provide evidence for the effects of resource availability on organisms and populations of organisms in an ecosystem. *MS-LS2-2. Construct an explanation that predicts patterns of interactions among organisms across multiple ecosystems.*	Interpretive • *Cite specific textual evidence to support analysis of science and technical texts (ELA)* • *Reading closely and explaining interpretations and ideas from readings (ELD)* Collaborative • *Interacting via written English (print and multimedia) (ELD)* Productive • *Write informational and explanatory texts (ELA/ELD)* • *Using nouns and noun phrases to expand ideas and provide more detail (ELD)*
Anchor Phenomenon	**Anchor Text**
Changes in sea otter, sea urchin, and kelp populations over time	"Ecosystem Superheroes: Sea Otters Help Keep Coastal Waters in Check" (*The Guardian,* adapted by Newsela staff, 11/14/2019)

Culminating Assessment Activity:

Ecologist Research, Blog, and Community Gathering

Targeted Language Demands

- **Language Functions:** analyze, interpret, explain
- **Vocabulary:** resource (recurso), ecosystem (ecosistema), population (población), community (comunidad), analyze (analizar), interpret (interpretar), biotic factor (factor biótico), abiotic factor (factor abiótico), Earth natural processes (Tierra procesos naturales)
- **Syntax:** Construct sentences that convey relationship and patterns found in data from a graph, including cause and effect relationships
- **Discourse:** Communicate an analysis and an interpretation of a phenomenon. Communicate an evidence-based explanation.

languages, and home/family/community funds of knowledge to build upon), standards (NGSS, CCSS ELA, and ELD), and the big science idea—the relationship between *how natural processes and human activities shape Earth's resources and ecosystems*. The frame also includes a naturally occurring event (or phenomenon), text (e.g., a science article), and culminating assessment activity that apply the big idea and standards in more relevant contexts. Finally, the unit frame includes discipline-specific language demands that students will need targeted support in to achieve both the NGSS and language and literacy standards. It is important to note that the unit was just one part of a larger module, Earth's Resources and Ecosystems, which would occur over several months.

You can think about the unit frame as a final product of unit-level planning. But, as will be illustrated in the chapters to come, this is a complex process. To scaffold the unit planning process, we developed four planning tools, listed below.

- Planning Tool 1: Knowing Your Emergent Bilinguals and Their Families
- Planning Tool 2: Unpacking the Next Generation Science Standards and Curricular Resources
- Planning Tool 3: Weaving Together ELA/ELD Standards and NGSS
- Planning Tool 4: Framing the Unit Through Rich and Relevant Phenomena, Texts, and Assessment

Each planning tool includes guiding questions and prompts in which to make and record planning considerations and decisions. Appendix C includes blank copies of each planning tool. Our goal in Chapters 4–7 is to deconstruct this unit frame, the importance of each component, and how to use each planning tool. Specifically, in Chapter 4, Planning Tool 1 will help illustrate how to gather knowledge about emergent bilinguals and their families. In Chapter 5, Planning Tool 2 will show how to unpack NGSS and curricular resources, which also helps us start making connections to language demands. In Chapter 6, Planning Tool 3 will illustrate how to weave in ELA and ELD standards. And in Chapter 7, Planning Tool 4 shows how to frame the unit through science-, language-, and literacy-rich phenomena, texts, and assessment.

PLANNING A CONCEPTUAL AND LINGUISTIC PROGRESSION

Once you have an overall framing for the unit, you can turn your focus to how to scaffold emergent bilinguals' content and language learning over time. By a *linguistic progression*, we mean scaffolding how emergent bilinguals use language for particular purposes when making meaning of texts (i.e., interpretive), making meaning with others (i.e., collaborative) and communicating meaning with texts (i.e., productive). Another three planning tools will guide you in moving from unit framing to a progression and specific learning activities that focus on student interaction with texts. Chapters 8–11 illustrate this part of the planning process using Planning Tools 5, 6, and 7. Planning Tool 5 is the conceptual and linguistic progression. Planning Tool 6 involves analyzing an anchor text. Finally, Planning Tool 7 helps unearth more discipline-specific language demands (i.e., vocabulary, syntax, and discourse) that emergent bilinguals would likely encounter to meet the NGSS and language and literacy expectations.

- Planning Tool 5: Conceptual and Linguistic Progression of Learning
- Planning Tool 6: Analyzing the Anchor Text
- Planning Tool 7: Unpacking the Unit's Science-Specific Language Demands

Ultimately, the unit frame and the progression of learning sets the stage for planning daily learning objectives, learning activities, and assessment activities (see Figure 3.2). We were intentional throughout the entire planning process to ensure that these activities tie back to the standards (both science and language/literacy) and the phenomenon, text, and culminating assessment anchoring the unit. Learning and assessment activities provide the opportunities for science, language, and literacy instruction for emergent bilinguals. But you must also weave in rich and relevant language support so that emergent bilinguals successfully navigate these opportunities.

In Chapters 9–11, we illustrate learning activities throughout the unit that provide these opportunities and support as they pertain to interpretive language (Chapter 9), collaborative language (Chapter 10), and productive

Figure 3.2. Ecosystem Interactions and Resources: Daily Objectives and Activities

Day (# of min)	Learning Objective: Students will be able to . . .	Learning and Assessment Activities
1 (80–90)	Notice important features of videos and graphs that connect to ecological interactions.	**Engage with Anchor Phenomenon— California Coast Kelp Forest Video and Population Dynamics Graph** • Record notices and wonderings from video. • Identify key features of a line graph. • Wonder what natural resources or human activity might affect ecosystems.

(continued)

Figure 3.2. (*continued*)

2 (80–90)	Describe the role of resources in an ecosystem. Compare and contrast abiotic and biotic interactions.	**Explore Local Ecosystem (Sonoma County, CA)—Landpaths Field Trip #1, Interactive Science Notebook Assignment** • Record notices, wonderings, it reminds me of. . . . • Identify *resources*. Explore what mineral, energy, water, and soil processes occur in this location as well as a beginning understanding of why. • Introduce and identify *abiotic factors* and *biotic factors*. • Describe 1 or more abiotic/biotic interactions and 1 biotic/biotic interaction occurring in this ecosystem.
3 (80–90)	Model how ecosystems are organized, including the biotic and abiotic resources available in them.	**Ecosystem Organization and Resources— Ecosystem Interactive Group Slides (Part 1)** • View kelp forest slide and video clip. • Discuss/respond to the following questions: » What resources are in the ecosystem? Which resources are biotic factors?/ abiotic factors? » Build levels of ecology on assigned interactive slide (organism → population → community → ecosystem) » Describe 1 abiotic/biotic interaction and 1 biotic/biotic interaction occurring in this ecosystem. *Student reference/support*: Savvas Segment 3 Workbook—Topic 7, Lesson 1, Living Things and the Environment
4–5 (160–180)	Cite specific evidence from a text (close reading) to explain patterns for how sea otter, kelp, and sea urchin populations interact and use resources in their ecosystem.	**Explore the Anchor Text—Analyzing Textual Evidence from the "Ecosystem Superheroes" Article** • Initial read and annotation of text. • Complete leveled inquiry questions. • Share analysis and interpretation of graph using textual evidence.

(*continued*)

Figure 3.2. (*continued*)

6 (80–90)	Analyze and interpret data in a population graph to predict patterns for how sea otter, kelp, and sea urchin populations interact and use resources in their ecosystem.	**Explore Another Local Ecosystem (Sonoma County, CA)—Landpaths Field Trip #2, Interactive Science Notebook Assignment** • Record notices, wonderings, it reminds me of. . . . • Define *interactions* and explore interactions in this ecosystem. • Predict how organisms compete and share resources. *How does understanding patterns in interactions and resources help us take better care of our land/ecosystems?*
7 (80–90)	Construct an explanation that describes how different organisms interact and use resources in ecosystems.	**Explain Ecosystem Interactions— Ecosystem Interactive Group Slides (Part 2)** • Identify the type of interaction on each slide (use TCI Graphic Organizer as needed). • Revisit graph from anchor phenomenon and text (deepen thinking). • What interactions occur in this ecosystem? How do these interactions affect other populations or resources in ecosystems? • Choose 1 interaction. Based on the patterns in this interaction, how can you design a solution to take better care of this organism and Earth's ecosystems/resources?
7–8 (160–180)	Identify and evaluate sources that describe a population and its resources in an ecosystem of choice.	**Ecosystem Research (Part 1/Blog Post Part 1)** • Students choose an ecosystem type and population/species living there to research (based on student/family interests). • Review how to evaluate sources and how to paraphrase/take high-quality research notes. • Students complete Part 1/Blog Post 1 research notes in Interactive Science Notebook.

(*continued*)

Figure 3.2. (continued)

9 (80–90)	Identify and evaluate sources that describe a population and its resources in an ecosystem of choice.	**Ecosystem Research (Part 2/Blog Post Part 2)** Students complete Part 2/Blog Post 2 research notes in Interactive Science Notebook.
10 (80–90)	Accurately record, graph, analyze and interpret population data in order to predict patterns of how organisms compete for and share resources in a California coast tide pool ecosystem.	**Analyzing Shell Beach Tide Pool Population Dynamics** • Land acknowldgement. • Data collection, analysis, and interpretation.
11–12 (160–180)	Construct an explanation in a published blog that predicts how different organisms interact and share resources in a variety of ecosystems on Earth.	**Evaluate/Culminating Assessment Activity** Ecologist Research and Blog Writing and Community Gathering.

language (Chapter 11). These chapters also include actual vignettes and student work to illustrate how it might play out in the classroom. We also provide several lesson plans in Appendix A to model lesson-level planning for you.

CONCLUDING REMINDERS

When planning instruction for emergent bilinguals, consider whether the unit and the more-detailed lessons and activities

- position all students as sense-makers, facilitate productive disciplinary discourse, scaffold language and disciplinary literacy, and contextualize instruction (i.e., the SSTELLA Framework);
- support emergent bilinguals in the interpretive, collaborative, and productive modes of communication; and
- support emergent bilinguals in developing and using discourse, syntax, and vocabulary/symbols for particular language functions (i.e., discipline-specific language demands).

These connections among the SSTELLA Framework, modes of communication, and language demands are recapped in Figure 3.3 below.

Figure 3.3. Connecting the SSTELLA Framework to Communication Modes and Language Demands

Overall, does the unit/lesson/activity . . .	
• position all students as sensemakers (support them in using a range of everyday intellectual and linguistic resources as they engage in science and/or engineering practices and make sense of science ideas)	• facilitate productive disciplinary discourse (talk moves, norms, structures, or resources to support students in exchanging and elaborating on information/ideas with others)
• scaffold language and disciplinary literacy (supporting interaction with texts and collaboration)	• contextualize instruction (frame instruction in contexts that are meaningful and relevant to students' lives in and out-of-school science)

AND support emergent bilinguals in the following modes of communication . . .
• Interpretive (making sense of texts)
• Collaborative (making meaning with others)
• Productive (communicating meaning with texts)

THROUGH the following discipline-specific language demands . . .	
• Language Function (overall purpose of using language)	• Discourse (*how* scientists or engineers talk, write, and reason about, for a particular purpose, i.e., *tools* that convey *meaning*)
• Syntax (how we *organize* words or symbols in science to convey meaning (i.e., *tools* that *organize*)	• Vocabulary/symbols (technical and nontechnical words, phrases, symbols to communicate precisely and efficiently

Part II

UNIT-LEVEL PLANNING

CHAPTER 4

Knowing Your Emergent Bilingual Students and Their Families

Now that we've shared the model unit, Ecosystem Interactions and Resources, in Chapter 3, we will begin discussing the process for planning a science unit that weaves in rich and relevant language support for emergent bilinguals. This process starts by gathering a rich set of relevant knowledge about your own emergent bilinguals and their families. Planning Tool 1 can assist you in fostering an asset-based mindset around what your students and families bring to your science learning community and, in turn, planning instruction and assessment that leverages what was learned. During the chapter, consider the following guiding questions:

- How can you gather a rich set of relevant knowledge about your own emergent bilinguals and their families?
- Why is this important, and how can you leverage what was learned while planning a science unit?

UNDERSTANDING EMERGENT BILINGUALS AS A DIVERSE STUDENT GROUP

Since emergent bilinguals do not reflect a homogenous group, teaching should not just take a one-size-fits-all approach when supporting them (Lee & Stephens, 2020). Beyond finding out the academic background (e.g., through transcripts, course history) and language proficiency (e.g., through state language development assessments) of all emergent and experienced bilinguals, you should collect an expanded set of knowledge about individual emergent bilinguals for the following critical reasons:

- *To avoid assumptions.* Especially if you have not experienced what it is like to learn a second language while learning science, it is important to avoid making assumptions about emergent bilinguals' abilities, experiences, and identities.

- *To value and leverage who they are.* Foster an asset-based mindset that views the lived experiences, linguistic repertoire, and home/family/community funds of knowledge as resources to leverage for planning or to adapt during instruction (including disciplinary connections). Instead of assuming that it is the norm to be monolingual and that emergent bilinguals are in "need of something," value multilingualism with all the assets our students and families contribute to our learning community.
- *To contextualize to emergent bilinguals' own lives and experiences.* One of the four dimensions of the SSTELLA Framework is to make science instruction meaningful and relevant (or contextualized to students' lives and experiences). Knowing about your emergent bilingual students' experiences allows you to frame an entire unit or specific activities in these relevant contexts.
- *To identify and plan targeted language support.* By collecting rich information about your emergent bilinguals' abilities, experiences, and interests related to language and literacy, you can better plan for targeted support. For instance, one emergent bilingual might be an avid reader in both their first language and developing in English, but struggle as a writer. Another emergent bilingual might be the exact opposite.

PLANNING TOOL 1: KNOWING YOUR EMERGENT BILINGUALS AND THEIR FAMILIES

Student and family ways of being and lived experiences should help drive curricular, instructional, and assessment decisions. Barton and Tan (2009) remind us that valuing nontraditional or nondominant ways of being in our learning spaces legitimizes multiple ways of participating and engaging in science. *Planning Tool 1: Knowing Your Emergent Bilinguals and Their Families* intentionally puts into practice this important paradigm shift. It aims to create an intentional practice around listening to and lifting up the stories and experiences of the students and families in our curriculum and learning experiences. What we ask students to do, what we engage them in, and how we assess them should meaningfully connect to their needs, interests, identities, and lives. Unit and lesson planning should not be a decision made *for* students and families but rather a dynamic process that happens *with* students and families throughout each unit. This is particularly important for our emergent bilingual students and families.

The planning tool is divided into three parts (see Figure 4.1): (1) Gather expanded relevant knowledge 2–4 weeks before the unit begins, (2) analyze the knowledge gathered, and (3) decide how to integrate what you learn about students and families into a unit.

Figure 4.1. Planning Tool 1: Knowing Your Emergent Bilinguals and Their Families

Guiding Questions

- How can you gather expanded relevant knowledge about emergent bilinguals and their families?
- How can you analyze and leverage this knowledge when planning your unit?

Part 1: Gather Expanded Relevant Knowledge (2–4 weeks before the unit begins).

Which methods will you use to gather knowledge about (1) personal interests and out-of-school experiences, (2) home/family/community funds of knowledge, and (3) ways to communicate in English and in languages other than English that are *relevant* to the science unit?

- ☐ Observe students read, write, and talk in class.
- ☐ Send out student and family survey related to the focal unit.
- ☐ Assign students to interview caregivers or other supportive adults.
- ☐ Focus on student and/or family/caregiver interviews.
- ☐ Family science night or other community event.

Part 2: Analyze Knowledge Gathered

What have you learned about students and families that is *relevant* to the science unit?

- **Personal interests and out-of-school experiences**
  - » Local parks, playing with and walking their dogs, playing basketball and soccer, biking/mountain biking, skateboarding, spending time with family and friends
  - » Visiting ocean/beach, local parks, and spending time in their garden
  - » Students enjoy learning when they have space to explore and discover outside
- **Home/family/community funds of knowledge**
  - » Students/families spend time outside in natural spaces to feel happy, peaceful, relaxed, calm, and spiritually connected. Water and trees are highly valued ecosystem resources. Spending quality time together with family and community is highly valued and often connected to experiences outside in nature, especially when visiting the ocean.
  - » Soccer, basketball and other sports are important ways of recreating and connecting together outside
  - » Water and trees as resources in our ecosystems
  - » Family spends time together in community
  - » Mindful moments in natural places outside where they can feel peace, calm, relaxed, happy, and spiritually connected
- **Ways to communicate in English and in languages other than English**
  - » **Primary and home language context**
    - o Oral communication in small or larger community groups
    - o Students communicate thinking and ideas with one another via Instagram, Snapchat, and other social media platforms outside the classroom
    - o 69% of families and students speak Spanish as their primary (or native) language
    - o 1 family/student speaks Fijian as their primary language

(continued)

Figure 4.1. (*continued*)

- o 1 family/student speaks Vietnamese
- o The remaining 35% of families speak English (or communicate in both English and Spanish)
- o Of the 35% (= 10) English-only students in this class, half of them attended Cesar Chavez Language Academy during elementary school years and received a dual immersion English/Spanish education TK–6
- o 31% of students in the class are designated English Learners with 1 newcomer student who has been in the United States for less than 3 years
- o 40% of students are designated Redesignated Fully English Proficient (RFEP) at this time
  - » **Oral language**
    - o Talking and interacting in spoken English and Spanish occurs with families during one-on-one phone conversations and in meetings as well as during interview assignments and school community events/meetings. These are always bilingual, with Spanish translation help available for monolingual English-speaking staff and Spanish-speaking families.
  - » **Written language**
    - o 69% of families read/write in Spanish. 1 family reads/writes in Fijian. 1 family reads/writes in Vietnamese.
    - o 35% of families read/write in English
    - o Some families and students are not fully able to read and write in Spanish or English
    - o Families use written English and Spanish when communicating via text messages and emails
    - o Students frequently use social media platforms (Instagram and Snapchat primarily) when reading and writing in both English and Spanish
    - o Students also use written English and Spanish in text messages, email communication, and school assignments

---

Part 3: Decide How to Leverage What You Learned About Students and Families Into a Unit

---

- **What are the most essential student and family assets that connect to the big idea or learning task of the unit?**
  - » Students and families are most interested in and connected to ocean/coastal and local Sonoma County park ecosystems.
  - » For families, the most important resources on Earth are water and trees because water is critically important for life and trees provide us with oxygen and clean air.
- **How can they be used to frame the unit?**
  - » Plan unit around ecosystem that is of interest to students.
  - » Take students to those places of interest (virtually or physically).
  - » Give students opportunities to conduct research (as an "ecologist") on organisms of their choosing to explore new places and consider solutions for better taking care of those organisms.
  - » Invite families and community members to a symposium at the end of the unit.
  - » Frequently allow students to use their multilingualism (English and Spanish, Fijian, or Vietnamese) in class.

Gather Expanded Relevant Knowledge

You might have gathered knowledge about emergent bilingual students and their families through records kept by the school about a student's native or primary language(s), when they first entered the county, their birth country, and their identified race/ethnicity. In California, for example, these records include data from English Language Proficiency Assessment (ELPAC) scores and state ELA, math, and science assessment scores, as well as academic transcripts that provide a historical view of a student's class schedules and grade performance. These are important, albeit limited, ways of knowing what language(s) might be being used by students/families, a student's country of origin, and what students have demonstrated they can do in our educational system. They do little to help you understand your students' authentic lives outside the classroom. To do this, you need to listen and learn from their stories, their experiences, and their beautifully rich humanity. In the planning tool we list multiple methods of gathering *expanded relevant knowledge* about emergent bilinguals'

- Personal interests and out-of-school lived experiences
- Home/family/community funds of knowledge
- Ways to communicate in English and in their home language

We suggest that you start with a family/community interview assignment with students and send out a family survey 2 to 4 weeks before each new unit begins. When planning and using these methods, consider the central scientific ideas or big science idea that students will be learning in order to understand how student and family assets may connect with these scientific ideas. For example, in the larger module on Earth's resources and ecosystems, we were particularly interested in what local or global outdoor ecosystems and resources were meaningful for students and families. We asked students to interview one important adult in their family or community and ask them questions about what places and resources in nature matter to that adult (see Figure 4.2).

We eventually organized opportunities for students and families to explore Earth's ecosystems and resources at local outdoor spaces with Landpaths, a nonprofit community conservation organization with the mission of fostering a love of the land in Sonoma County (see www.landpaths.org). We recommend investigating what local outdoor or real-world learning opportunities are available for students in your area to explore their local resources, ecosystems and environments, and science careers. This can include city, regional, state, or national park systems and nonprofit environmental agencies focused on connecting students and families to outdoor spaces with barriers of cost and transportation being thoughtfully addressed. In particular, we

Figure 4.2. Community/Family Interview

Earth's Resources and Ecosystems Community/Family Interview -
What Places and Resources in Nature Matter to You?

Directions:
BEFORE the INTERVIEW:
- Choose **one or more adults** in your community or family to interview.

- Decide and prepare for how you will capture important ideas from the conversation.
    a. *Options include:*
        - Hand-write or type shared ideas
        - audio-record (app on your phone or device or...)
        - video-record/Zoom session record
        ***Before audio or video recording, be sure to ask permission of the adult you are interviewing.

DURING THE INTERVIEW:
- Find a **quiet place** where you can talk and share ideas for about 30 minutes or more together.

- Ask your Interviewee each of the Interview Questions below. After each question, pause, listen carefully and record/take notes on their responses for each question.

Interview Questions:
1. What is one of your **favorite outdoor places in nature** to visit? Why?

2. How do you feel when you are at this (or these) places in nature? Why?

3. What is the **most special plant, animal, or other organism** you have experienced in nature? Why was or is it so special?

4. What is a **resource** on Earth that you think is very important? Why?

(**resource** - any material or form of energy needed by living things to survive, grow and reproduce *Examples: air, water, soil, fossil fuels, sunlight, wind, food, minerals, space*)

---

recommend looking for organizations that value and lift indigenous wisdom and stewardship as well as multicultural wealth in our communities.

We also sent families a Google Form Survey that included information about Landpaths and asked, "What Landpaths Outdoor Locations/ Preserves in Sonoma County would you like your student or family to visit?" (*¿Qué lugares/reservas de Landpaths Outdoor en el condado de Sonoma le gustaría que su estudiante o su familia visitaran más?*) During the unit, we also used our new district communication tool, Parent Square, to communicate to students and families about what complex learning inquiries and tasks would be part of the unit, what essential terms and big science ideas students would learn, and what opportunities would exist to coconstruct the learning with students and families.

Finally, throughout each unit, we recommend planning for intentional opportunities for students to express their ideas in both written and oral communication to gather a deeper understanding of the language assets they bring into our learning communities. This can begin with many short, formative assessment opportunities that include daily warm-up quick writes, pair-share and small-group team conversations, whiteboard brainstorms, online discussion boards, Google Form short answer responses, and interactive notebook written responses. These will allow the teacher to cocreate learning activities, assessments, and supports based on the language, experience, and communication strengths and interests of the students and families in their context.

Analyze Knowledge Gathered

We provided an extensive list of knowledge gathered in the sample planning tool. Just the interview alone was rich in information. Students chose to interview a wide variety of adults in their community including their mom, dad, both parents, aunts, cousins, older brothers or sisters, a friend, their grandparent, and one student chose to interview himself (as everyone else was too busy).

Overall, the ocean and redwood forests were shared most often as their adult's favorite outdoor place to visit in nature. Many shared that they enjoyed the fresh air, feeling sand between their toes, and hearing moving water. Being in these locations made them feel peaceful, calm, and happy. In addition, many families shared that oak woodland local parks were favorites that helped them experience similar mindful moments of peace, calm, and happiness. A few adults shared locations where they had grown up in Colorado and Mexico. One student shared that she wanted to know more about her grandma's childhood in El Salvador after the interview. Most of the adults interviewed shared that water and trees were the most important resources in our ecosystems. The common theme was that trees provide oxygen and shelter, and help clean our air. For families and adults in our communities, water is life and precious to our Earth and its ecosystems.

We share these details to highlight the rich source of knowledge about students and families that are not captured on language tests and are sometimes difficult to build upon spontaneously if shared in class. Every student will have relevant knowledge to share that differs from one another. In this way, the science teacher is already differentiating instruction by understanding the diversity within their classroom.

Some students did struggle to find an adult to interview, communicating with us that adults in their lives were inaccessible due to hours of employment, conflicts in their relationship, and/or language barriers. For these students, we personally interviewed them to learn more of their experiences and values. We also invited students to respond to questions on their own from their own perspective, to interview peers, and/or to interview another staff member who is important to them. Integrating these interview assignments consistently into each unit provides all of us—teachers, students, and families—multiple touchpoints to overcome barriers as we learn and coconstruct learning together. In addition, consistently using multiple methods (e.g., surveys, interview assignments, learning circle, focus group interviews) to learn more about our students and families before, during, and/or after each science unit also provides more opportunities to grow in our understanding of all the assets our students and families bring into our learning communities.

Decide How to Leverage Knowledge Learned

When looking across all data sources, students and families were clearly most interested in and connected to ocean/coastal and local Sonoma County park ecosystems. Families noted how water and trees are the most important resources on Earth because water is critically important for life and trees provide us with oxygen and clean air. While it was known that Spanish was the predominant, but not only, language spoken at home, we uncovered the variety of oral and written forms of communication used. We also uncovered the important connections students and families have with their community.

These essential student and family assets helped us frame the science unit. We decided to plan a unit around an ecosystem that is of interest to students—the kelp forest. We took students to a place of interest that was relatively close (within an hour's drive): Shell Beach. Shell Beach was chosen because it was a beach location that was open at no cost for students and their families to visit during the pandemic, with accessible tidepooling for students and families to experience the organisms in our lesson. Shell Beach is also a location where Landpaths takes students during spring break and summer camps that were made available to many of our students at low or no cost during this school year. Of course, due to the COVID pandemic, Landpaths arranged for us to take a virtual field trip.

We also have student opportunities to conduct research (as an "ecologist") on organisms of their choosing, explore new places, and consider solutions for taking better care of those organisms. Finally, we invited families and community members to a gathering at the end of the unit where students would share and discuss their organism and ecosystem research. Throughout the unit, we encouraged students to use their multilingualism (English and Spanish, Fijian, or Vietnamese) in class. We also communicated with families continuously through a weekly communication blast that (1) shared what their students would be doing/learning in the upcoming lessons, (2) acknowledged and celebrated student ideas and assets learned, and (3) invited feedback/input from families and students.

In the words of Barton and Tan (2009), these hybrid spaces are "moments where science is no longer a separate world" from students' lives outside of the classroom. They are *supportive scaffolds* that link traditionally marginalized ways of being to academic ways of being so that students can demonstrate proficiency in science learning and apply their science learning

in their communities and everyday lives. Now, you need to couple this understanding of your student context with an understanding of your curricular context—the focus of Chapter 5.

CONCLUDING REMINDERS

- Get to know your emergent bilinguals and their families to avoid
  assumptions, value and leverage who they are, contextualize
  instruction, and plan targeted language support.
- Gather and analyze knowledge about (1) personal interests and
  out-of-school experiences, (2) home/family/community funds
  of knowledge, and (3) ways to communicate in English and in
  languages other than English that are *relevant* to the science unit.
- Leverage this expanded set of knowledge to intentionally create
  spaces where emergent bilinguals and families are lifted as active
  creators of their science learning.

CHAPTER 5

Unpacking the Next Generation Science Standards and Curricular Resources

This chapter focuses on knowing the curricular context for your course, including Next Generation Science Standards and other resources such as adopted textbooks and curriculum, key investigations, articles, and online materials. We describe how to use Planning Tool 2: Unpacking the Next Generation Science Standards and Curricular Resources to (1) identify essential NGSS performance expectations, (2) arrive at a big science idea statement, and (3) begin identifying key discipline-specific language demands—namely, vocabulary and language functions. For this chapter, consider the following guiding questions:

- How can you unpack the conceptual ideas listed in Next Generation Science Standards and curricular resources?
- How do science and engineering practices help you identify key ways in which students will be using language?

UNDERSTANDING YOUR CURRICULAR CONTEXT

Alongside gathering information about emergent bilinguals and their families, you need to deeply understand what you plan to teach and the resources available to design and sequence learning activities and assessments. Science standards provide you with learning (or performance) expectations (i.e., what students should know and be able to do), but they also help guide some of the ways students would need to use language to make sense of and communicate those expectations. Science and engineering practices, in particular, will help you identify the primary functions of language as they relate to specific performance expectations. You also need to carefully unpack additional curricular resources to unveil the most central disciplinary idea (or what we call a "big science idea") that emerges when looking across science ideas addressed in these different resources.

PLANNING TOOL 2: UNPACKING THE NEXT GENERATION SCIENCE STANDARDS AND CURRICULAR RESOURCES

Planning Tool 2 is organized around two parts: (1) identifying a set of essential NGSS performance expectations (PEs) and (2) unpacking standards and curricular resources to arrive at a big science idea statement, essential vocabulary, and focal language functions. By completing Planning Tool 2, you will be in a better position to not only know what science you will be teaching, but also how to weave in language and literacy standards—the focus of Chapter 5. (See Figure 5.1.)

Figure 5.1. Planning Tool 2: Unpacking the Next Generation Science Standards and Curricular Resources

Guiding Questions
• What are the essential NGSS performance expectations? • How can you unpack standards and curricular resources to arrive at a big science idea statement, key vocabulary, and language function for the unit?

Unit Title: *Ecosystem Interactions and Resources*

Part 1: Essential NGSS Performance Expectations

Essential Performance Expectations of the Unit (3–5 max)
MS-LS2-1. Analyze and interpret data to provide evidence for the effects of resource availability on organisms and populations of organisms in an ecosystem. *MS-LS2-2. Construct an explanation that predicts patterns of interactions among organisms across multiple ecosystems.*

Focal **Disciplinary Core Idea(s) (DCI)**	Life Science Core Ideas • *LS2.A: Interdependent relationships in ecosystems* Earth and Space Science Core Ideas • *ESS3.A: Natural resources*
Focal **Science and Engineering Practice(s) (SEP)**	• *Analyzing and interpreting data (to provide evidence for phenomenon)* • *Constructing explanations and designing solutions (that includes qualitative or quantitative relationships between variables that predict phenomena)*
Focal **Crosscutting Concept(s) (CCC)**	• *Cause and effect* • *Patterns* • *Structure and function*

(continued)

Figure 5.1. (continued)

Part 2: Unpacking Standards and Curricular Resources to Arrive at a Big Science Idea Statement, Key Vocabulary, and Focal Language Functions

Resources Reviewed	• *NGSS* • *California Board of Education 2016 Science Framework* • *TCI Bring Science Alive! (7th-Grade Integrated Science)* • *Savaas Elevate Science (California Integrated Grade 7)*

Key Concepts and the Big Science Idea

Savvas Segment 3 California Integrated Integrated Grade 7 - ANCHORING PHENOMENON:

How does the **human consumption of natural resources** affect **ecosystems**?
- Topic 6: Distribution of **Natural Resources**
- Topic 7: **Ecosystems**

CA Science Framework
Grade 7 Integrated Storyline GUIDING CONCEPT:
Natural processes and **human activities** cause **energy** to flow and **matter** to cycle through *Earth systems.*

(CA Board of Education 2016 Science Framework, pg. 422)

Big Science Idea:
Natural processes and human activity affect Earth's resources and ecosystem

TCI Instructional Segment 3 -
The Distribution of Earth's Resources
Processes that Shape Earth
- Lesson 16 - Earth's Tectonic Plates
- Lesson 17 - The Rock Cycle
- Lesson 18 - The Water Cycle
- Lesson 19 - **Earth's Natural Resources**
Resources in Ecosystems
- Lesson 20 - **Resources** in Living Systems
- Lesson 21 - **Interactions** Among Organisms
- Lesson 22 - Changing **Ecosystems**

CA Science Framework
Grade 7 Integrated Storyline SEGMENT 3:
Natural processes and **human activities** have shaped Earth's **resources** and **ecosystems**.
- Life Science Storyline: **Resource availability** affects organisms and ecosystem populations.
- Earth Science Storyline: Geoscience processes **unevenly distribute** Earth's mineral, energy, and groundwater resources.
- Physical Sci Storyline: Chemical reactions make new substances Mass is conserved in physical changes and chemical reactions

(CA Board of Education 2016 Science Framework, pg. 422)

Key Supporting Vocabulary	Core Idea-related	SEP-related:	CCC-related:
	• *resource* • *habitat* • *ecosystem* • *interdependence* • *relationship* • *interaction* • *natural process* • *human activity*	• *analyze* • *interpret* • *evaluate* • *explain*	• *patterns* • *cause* • *effect*
Focal Language Functions	*Analyze, Interpret, Explain*		

Essential NGSS Performance Expectations

We propose starting to hone in on essential NGSS performance expectations by identifying the NGSS disciplinary core ideas most appropriate for the grade level, course scope and goals, and timing within the course. As described in Chapter 3, we situated our unit within the second semester of a middle school 7th-grade integrated science course. This course wove together the physical sciences, life sciences, and earth and space sciences in each unit. In the *2016 California Science Framework for Public Schools* (see website at https://www.cde.ca.gov/ci/sc/cf/cascienceframework2016.asp), NGSS integrated segment 3 suggests six disciplinary core ideas. You may be working from adopted curriculum or district/school pacing guides that already lay out core ideas. It is still good to understand the rationale for performance expectations chosen for a particular adopted curriculum or pacing guide.

The six disciplinary core ideas from integrated segment 3 of the *2016 California Science Framework for Public Schools* included 12 performance expectations. Looking across the PEs, we noticed how the ideas of "ecosystems" and "resources" related to and connected multiple PEs. For that reason, the larger module was titled *Earth's Ecosystems and Resources*. It is a daunting task to plan a unit around so many PEs. Therefore, we suggest narrowing the list of all PEs down to 1–3 "essential" PEs needed to address the primary disciplinary core idea (or two) for a 2–4-week unit.

When looking deeper at all three dimensions (i.e., core ideas, crosscutting concepts, and science or engineering practices) of each PE, you can begin to notice which PEs might be at the heart of the whole unit and which ones might just play a supporting role. For us, the crosscutting concepts revealed a focus on "pattern" and "cause and effect" while science practices focused on "analyzing and interpreting data" and "constructing explanations." (See the example in Figure 5.2.) Thus, we divided the larger Earth's Ecosystems and Resources module into several units, including Ecosystem Interactions and Resources. Ecosystem Interactions and Resources focused on the two essential performance expectations displayed in Planning Tool 2.

Figure 5.2. Example of Essential NGSS Performance Expectations and Associated Science and Engineering Practices and Crosscutting Concept

MS-LS2-1. Analyze and interpret data to provide evidence for the effects of resource availability on organisms and populations of organisms in an ecosystem.

Disciplinary Core Idea	Crosscutting Concept	Science and Engineering Practice (SEP)
LS2.A: Interdependent relationships in ecosystems	Cause and effect	Analyzing and interpreting data (to provide evidence for phenomenon)

Big Science Ideas

Windschitl et al. (2018) included planning for engagement with big science ideas as part of their "coherent and accessible vision of what highly effective science instruction can look like" (p. 3). Put more simply, they referred to their vision as ambitious science teaching. A big science idea expresses relationships, not facts, to connect and *explain* a range of other ideas. NGSS uses a similar term, *disciplinary core idea* (e.g., "Interdependent relationships in ecosystems"). Disciplinary core ideas are also meant to coalesce many science concepts, principles, and theories. As written, disciplinary core ideas vary in how well they *explain* other ideas. Also, you are likely to pull from multiple curricular resources, not just NGSS, to plan instruction that includes their own concepts and learning goals. It is important to unpack all key resources to arrive at a singular big idea that can unify and guide planning. A big science idea might focus on the NGSS disciplinary core idea, clarify it, and/or provide a clearer sense of the relationships and connections across concepts. Science teachers we worked with talk about how big science ideas, when related back to students' lived experiences, help students know why they should care about learning these concepts.

To arrive at a big science idea, we suggest starting with language from the essential NGSS performance standards and/or state science standards/frameworks. Read through them and then write down key concepts/ideas/science practices (or components thereof) that appear. This is a good exercise to also understand what students are encountering. Then do the same for the adopted textbook and/or curriculum. You can also consider key investigations or projects typically used by you and/or your colleagues. By looking within and across each resource you can then generate a visual representation of how concepts are related to each other, such as the spider diagram in the middle in Figure 5.1 or concept map. A *spider diagram* helps show related concepts and the more central concept. We have found that 3–5 central concepts often emerge in a unit. We show this as the four spokes of our spider diagram. Each spoke includes a question or statement, such as "The Distribution of Earth's Resources."

From the central concepts, you can construct a clear and succinct statement (not a question) that explains a range of science ideas. In our unpacking, it became clear that the concept of *resources* was central and related to all areas of the unit. It is a concept broad enough and deeply enough connected to all three disciplines of physical, life, and earth science to allow for many entry points in the unit. Students have the opportunity to explore what resources occur in different areas of our local and global community and ecosystems, why they occur in these specific locations, how they are important to populations in ecosystems, and how they are impacted by human activity and natural earth processes. The relationships between the four big ideas of this unit became clear in this analysis—*resources, ecosystems, natural processes,* and *human activity.* Many interesting and contextualized learning opportunities can be made available and coconstructed with students and families with these interrelated big ideas. The words *affect, shape,* and *influence* also frequently appeared in this analysis, which signaled the importance of cause–effect relationships in these science ideas that students would explore. Thus we eventually constructed the big idea statement *Natural processes and human activity affect Earth's resources and ecosystem.*

Essential Supporting Vocabulary

Key concepts can be further unpacked to identify vocabulary related to the disciplinary core ideas, the crosscutting concepts, and the science or engineering practices. Here lies an important point about how we will think about vocabulary throughout this book. Students, including emergent bilinguals, do not need to learn all science vocabulary first to make sense of a concept or idea fully. All students enter the classroom with linguistic resources that can assist in beginning to understand photosynthesis, Newtonian mechanics, or plate tectonics. But students are learning and using vocabulary, syntax, and discourse in the service of more precisely explaining and analyzing these complex and abstract concepts and ideas that will allow them to deepen understanding and apply them as they emulate the intellectual activities of scientists and engineers. For example, in a writing activity early on in a unit around meiosis and genetic diversity, a student might write: "*Also sexual reproduction is genetically unique. This way an species don't have the same skills & this gives a specie a more likely chance to survive.*" The word "skills" might be part of the student's linguistic repertoire and useful for beginning to connect genetic variation to the concepts of "natural selection" and "adaptation." Not only is "skills" a more commonly used word than "adaptation," but it might be used in other contexts, such as in sports, arts, games, cooking, or car mechanics, that could eventually bridge to the nuanced situation of adaptations. The key is to allow students to use more of their own vocabulary at first to begin developing conceptual understanding, then to support students' gradual development of related vocabulary to communicate ideas (such as in an argument) more clearly and precisely. This approach to teaching content before vocabulary is supported by research (e.g., Brown & Ryoo, 2008).

Essential Language Functions

In Chapter 2 we discussed considering how students will be actively using language for a particular purpose (i.e., a language function). In a way, the language function is the linguistic analog of the big idea. We identify a big idea to bridge together various concepts and know what we ultimately aim for students to understand. Similarly, honing in on the language function better allows us to identify how to support students in understanding and using particular features of language (e.g., vocabulary, syntax, and discourse) in different modes (e.g., making sense of written language or producing written language).

The simplest way to arrive at a language function is to look at the science and engineering practices for the unit, which start with the active uses of language—to "analyze and interpret" and to "explain." We could additionally look toward the crosscutting concepts for more nuanced use, such as interpreting patterns in data or to explain cause-and-effect relationships. But for our purposes, "Analyze," "Interpret," and "Explain" are three essential uses of language that can help guide us when looking closer at English language arts and English language development standards, described in the next chapter.

CONCLUDING REMINDERS

- Unpack available standards and curricular resources to arrive at a singular big science idea statement that can frame the whole unit.
- Examine the language of science or engineering practices and the language of crosscutting concepts carefully to identify what students will primarily be actively doing with language (i.e., the language functions).
- Collectively, unpacking NGSS and curricular resources ensures that emergent bilinguals have the opportunity for rich science and can better become aware of how science ideas are interconnected and call on them to also use language in rich ways.

Weaving Common Core English Language Arts and English Language Development Into Next Generation Science Standards

Chapter 6 discusses how to intentionally weave Common Core English Language Arts (CCSS ELA) standards and English language development (ELD) standards into NGSS. We begin by clarifying exactly what Common Core ELA and ELD standards are and how they are structured. Then we describe Planning Tool 3: Weaving Together ELA/ELD Standards and NGSS to show you how to strategically target ELA and ELD standards for a unit that has already been initially planned around essential NGSS performance expectations. While reading this chapter, consider the following guiding questions:

- What are the similarities and differences between Common Core ELA Standards and English language development standards?
- How do I choose language and literacy standards for a unit?

COMMON CORE ENGLISH LANGUAGE ARTS (ELA) STANDARDS

Common Core State Standards, released in 2010, refers to two sets of standards, one for English language arts (ELA) and one for math (see website at http://www.corestandards.org/). We focus exclusively on Common Core ELA in this book. Common Core ELA draws on the principle that by the time *all* students graduate from high school, they . . .

- build strong content knowledge.
- respond to the varying demands of audience, task, purpose, and discipline.
- comprehend as well as critique.
- value evidence.

- use technology and digital media strategically and capably.
- come to understand other perspectives and cultures.

This list represents the practices of literate individuals. But even in this list we can identify some parallels with NGSS principles, such as the continued emphasis on content knowledge (i.e., disciplinary core ideas) and the new emphasis on the science practices of explaining and arguing with evidence. To articulate expectations for these literacy practices, Common Core ELA is organized around four anchor standards: *Reading, Writing, Speaking,* and *Listening,* in addition to *Language: Conventions, Use, Vocabulary.* Common Core ELA is divided into two sections for Grade 6–12 teaching—one for the ELA teacher and one for the science and history teachers labeled "Grades 6–12 Literacy in History/Social Studies, Science, & Technical Subjects." This division can reduce how overwhelming it would feel as a science teacher to tackle ELA in addition to science standards. The job becomes even less daunting, and in many ways complementary, when looking closely at what is expected of Grade 6–12 science teachers.

The Common Core ELA section "Grades 6–12 Literacy in History/ Social Studies, Science, & Technical Subjects" only includes standards related to reading and writing (not listening, speaking, and language conventions). The Science & Technical Subjects Reading standards are distinct from the History/Social Studies Reading Standards, whereas the Writing standards are the same for Science/Technical Subjects and History. These 20 Common Core ELA standards (10 reading, 10 writing) are organized around the following categories:

Reading

- Key Ideas and Details (Standards 1–3)
- Craft and Structure (Standards 4–6)
- Integration of Knowledge and Ideas (Standards 7–9)
- Range of Reading and Level of Text Complexity (Standard 10)

Writing

- Text Types and Purposes: Write Arguments (Standard 1)
- Text Types and Purposes: Write Informative/Explanatory Texts (Standard 2)
- Production and Distribution of Writing (Standards 4–6)
- Research to Build and Present Knowledge (Standard 7–9)
- Range of Writing (Standard 10)

Now let us turn to the purpose and structure of commonly used ELD standards and frameworks.

ENGLISH LANGUAGE DEVELOPMENT (ELD) STANDARDS

What exactly does "English language development" mean, and how does it relate to English language arts and learning science? In the United States, while Common Core ELA lays out what it means for all students, including emergent bilinguals, to become literate and college- and career-ready (e.g., comprehend as well as critique, value evidence), ELD refers to targeted capabilities that support the emerging use of the English language to do academic work. It may help to think about language as a tool for communicating ideas. So ELD is primarily about developing students' linguistic repertoire (in English) so that they are supported in learning science and other content areas.

While ELD standards and this book focus more on *English* language development, it is also critical for you to provide spaces for emergent bilinguals to read, write, and talk in all languages available to them. The purpose is not just to make science content more accessible to emergent bilinguals (such as by translating lab instructions), but rather to leverage emergent bilinguals' primary language, along with their emerging English, as tools to make deeper meaning of complex science ideas. For decades, research has documented the benefits for all students to develop bi- or multiliteracy, but there has been increasing awareness of how biliteracy, not just "English" language development, can be supported in mainstream, English-dominant science classrooms (e.g., Suárez, 2020). In Chapters 8–11, we provide examples of how mono- or bilingual science teachers can help emergent bilinguals progress by using all available linguistic resources to explore phenomena, interpret data, and produce evidence-based explanations.

Across the United States, the WIDA English Language Proficiency Standards (see https://wida.wisc.edu/sites/default/files/resource/2012-ELD -Standards.pdf) and the English Language Proficiency Assessment for the 21st Century (ELPA21) Standards (see https://elpa21.org/wp-content /uploads/2019/03/ELPA21-Organization_of_Standards-5.22.15.pdf) are the most widely adopted sets of ELD standards in the United States. California, the state in which our model unit was developed, developed its own California ELA/ELD Framework that integrates Common Core ELA with previously adopted ELD standards (California Department of Education, 2015). The ideas and planning tools used throughout this book can be adapted for use with any set of ELD standards. However, we will use the California ELA/ ELD Framework given the geographic context and how this framework reflects the view of deeply integrating ELA and ELD. The California ELD Standards are organized into three parts:

1. Interacting in Meaningful Ways
2. Learning About How English Works
3. Using Foundational Literacy Skills

ELD is not about learning the grammatical rules of the English language. Even Part 2 (Learning About How English Works) focuses more on how to structure cohesive texts, how to add details and enrich ideas, and how to connect and condense ideas. But as seen from the list, ELD is even more than just learning the structure of language; it is applying language structure to make meaning for particular *purposes*. And meaning-making involves interaction—both with texts and with other people. This book focuses most directly on planning instruction to support Interacting in Meaningful Ways, which is further divided into the following three modes of communication:

- *Interpretive language*: making sense of oral and/or written language. We will focus mostly on interpreting written language, which also includes visual models, charts, graphs, and other forms of representational texts.
- *Collaborative language*: communicating *with*, not just to, others to make meaning. Collaborative communication can happen orally (in English, first language, or strategically and fluidly between the two), with gestures and manipulatives, and through multimedia texts (e.g., Zoom Chat, Google Documents, and Jamboards) that have become more commonplace in schools since the COVID pandemic.
- *Productive language*: communicating meaning through oral and/or written language. We will focus mostly on the production of written language.

Finally, Part 3 (Using Foundational Literacy Skills) reminds teachers that they need to consider who the students are before and during use of ELD Standards to support them, as was discussed in Chapter 4.

There is an important and close link between Common Core ELA and ELD. The Common Core English Language Proficiency Standards with Correspondences to K–12 English Language Arts (ELA), Mathematics, and Science Practices, K–12 ELA Standards, and 6–12 Literacy Standards "highlight and amplify the critical language, knowledge about language, and skills using language that are in college-and-career-ready standards and that are necessary for English language learners (ELLs) to be successful in schools" (Council of Chief State School Officers, 2013, p. 1). Even though emergent bilinguals will produce language that includes features that *distinguish* them from their native-English-speaking peers, "it is possible [for emergent bilinguals] to achieve the standards for college-and-career readiness" (National Governors Association Center for Best Practices & Council of Chief State School Officers, 2010, p. 1). Similar to the Common Core literacy standards in Science/Technical Subjects, ELD standards connect language development with learning *content* "to support the dual aims of ensuring that all . . . [students] have access to intellectually rich academic content across the disciplines and that they simultaneously develop academic English" (California Department of Education, 2015). Emergent bilinguals who are not redesignated as English proficient yet can be supported through protected time outside of the content-area classroom by qualified ELD instructors (i.e., Designated ELD). However, this book is intended for science teachers who either teach in a mainstream science class with students reflecting various English language proficiencies or who teach in a sheltered class exclusively with emergent bilinguals. In this setting, the science teacher is engaging in integrated ELD.

PLANNING TOOL 3: WEAVING TOGETHER ELA/ELD STANDARDS AND NGSS

The ultimate goal of Planning Tool 3 (see Figure 6.1) is to intentionally target particular ELA/ELD standards for a given unit so that the standards seamlessly weave into the previously identified NGSS performance expectations, thereby creating a synergistic and reciprocal relationship between science learning and language/literacy development. The planning tool will also aid you when further unpacking discipline-specific language demands (i.e.,

Figure 6.1. Planning Tool 3: Weaving Together ELA/ELD Standards and NGSS

Guiding Questions

• What interpretive, collaborative, and productive ELA/ELD standards will best support science learning and language/literacy development throughout the unit?
• What additional opportunity is there to support how English works?

Interpretive

ELA Reading in Science: *Cite specific textual evidence to support analysis of science and technical texts*

ELD Interacting in Meaningful Ways: *Reading closely and explaining interpretations and ideas from readings*

Collaborative

ELD Interacting in Meaningful Ways: *Interacting via written English (print and multimedia)*

Productive

ELA Writing in Science: *Write informational and explanatory texts*

ELD Interacting in Meaningful Ways: *Writing literacy and informational texts*

Learning About How English Works

Structuring Cohesive Texts, Expanding or Enriching Ideas, Connecting and Condensing Ideas: Using nouns and noun phrases to expand ideas and provide more detail vocabulary, syntax, discourse) needed to interpret, collaborate, and produce. Together with Planning Tool 1: Knowing Your Emergent Bilinguals and Their Families, you have the foundational context in which to begin making standards relevant for students.

Starting With Language Functions

How do you refrain from *adding* ELA/ELD to the curriculum and instead *leverage* ELA/ELD as opportunities to simultaneously support emergent bilinguals' science learning and language and literacy development? We approach ELA/ELD planning intentionally. Where in the NGSS and curricular resources would it benefit students to develop deeper language and literacy development? And vice versa, which ELA/ELD standards would the planned NGSS performance expectations be able to support? Furthermore, what have we learned about our emergent bilinguals that helps us plan to offer support in areas of language and literacy where they most need it?

To begin, you can look at the language function that was identified in the last chapter after unpacking NGSS and curricular resources. Our model unit focused on the expectation for all students to "*construct an explanation that predicts patterns of interactions among organisms across multiple ecosystems.*" From the associated science practices of "Constructing an Explanation" and "Analyzing and Interpreting Data," we identified the two language functions: explain and analyze. Language functions serve as a key connection between NGSS and ELA/ELD standards because science content becomes the *context* in which language is practiced and mastered.

You can then look at our list of 20 Common Core ELA Literacy in Science and Technical Subjects Standards and the list of ELD standards across three modes of communication—interpretive language, collaborative language, and productive language. When choosing, we recommend being judicious. You might find many ELA/ELD standards that relate to the NGSS and language functions. However, it's important to identify a few standards for which you can *commit* to provide targeted and ongoing support during the unit.

Interpretive Language Standards

To focus on making sense of oral and/or written language (i.e., interpretive language), we reviewed the 10 ELA Literacy in Science and Technical Subjects: Reading Standards at the Grades 6–8 level. The following appeared to be a good match for our language function "analyze."

- *RST.6-8.1 Cite specific textual evidence to support analysis of science and technical texts.*

Students will be expected to draw on a range of capabilities to comprehend and analyze written and spoken texts. But because Ecosystem Interactions and Resources will rely heavily on a central text, we decided to also target the closely matched ELD standard *"Reading closely and explaining interpretations and ideas from readings"* to help emergent bilinguals learn to navigate the text to make sense of the content, practice of science, and evidence that will help support future learning.

Collaborative Language Standards

There are many opportunities for collaboration in science classrooms, such as when students pair up to share responses to a warm-up question, work in small groups during an investigation, or give feedback to each other on lab reports, graphs, or models. The purpose of ELD is to expand the linguistic repertoire with which emergent bilinguals can successfully collaborate with others. For Ecosystem Interactions and Resources, we chose to focus on multimedia collaboration given how the COVID pandemic called upon students to interact through multiple multimedia platforms. The specific California ELD standard we targeted is *"Interacting via written English (print and multimedia)."* We imagined that these collaborative digital platforms would support emergent bilinguals, for one, in analyzing data to help write an explanatory text. Even print handouts can be used to support collaborative language. For instance, in a different unit, a simple handout was provided to students to summarize claims made by a partner around an article about whiptail lizards and their unusual form of reproduction. When summarizing claims made by a partner, emergent bilinguals can practice exchanging ideas through discipline-specific writing (note the student's use of "if . . . then" to reason scientifically). One advantage of this type of handout and activity is that they have the time to process and record thoughts, which may be a useful scaffold before more intensive back-and-forth verbal exchanges.

Collaborative language is not explicitly called out in CCSS ELA Literacy in Science and Technical Subject Standards, which focus on the fundamental modes of literacy—reading and writing. For this reason, it is essential to go beyond CCSS ELA and utilize ELD standards to guide science planning so that emergent bilinguals can expand their linguistic repertoire.

Productive Language Standards

To align with the language function "explain," we chose to focus on the standard WHST.6-8.2: *Write informative/explanatory texts to examine a topic and convey ideas, concepts, and information through the selection, organization, and analysis of relevant content.* We imagined that by the end of the unit students would be able to explain, with evidence from various learning activities, how natural processes and human activities influence Earth's eco-systems. To be able to effectively write this explanatory text, students need to deeply understand the core idea as well as the purpose, audience, and basic architecture of a scientific explanation. Thus, throughout the unit, learning activities would scaffold students' ability to write an explanatory text. For instance, early on they may be able to identify "causes" and notice patterns that reflect "effects." Later in the unit they might then practice communicating a cause-and-effect *relationship* visually (e.g., a flow chart) and in writing. A California ELD standard matches the ELA standard almost verbatim: *Writing literacy and informational texts.*

We can start seeing how instead of an additive process, there are opportunities to truly weave together ELA, ELD, and NGSS. So, while in the unit we will give emergent bilinguals the opportunities to write an informational text (in the form of an evidence-based explanation), we can support them through multiple lenses—a disciplinary lens (writing to convey science ideas in ways that resemble how scientists would convey them) and a language/literacy lens (attending to the structure, audience, and purpose of text to convey ideas).

Learning About How English Works Standards

Targeting interpretive, collaborative, and productive language standards can help emergent bilinguals make meaning of and communicate science ideas. You can additionally support meaning-making and communication for emergent bilinguals by helping them learn how to structure cohesive texts and how to expand, enrich, connect, and condense ideas. Just as students learn science vocabulary to more precisely communicate about abstract science ideas (not just for the sake of memorizing vocabulary), it is important for emergent bilinguals to know, for example, how to use "nouns and noun phrases" to expand ideas and provide more detail when communicating (rather than just memorizing language conventions).

DIFFERENTIATING ELA/ELD STANDARDS BY KNOWING YOUR EMERGENT BILINGUALS

Up to the point, we used the curricular context, through NGSS Performance Expectations, to guide the targeted and intentional selection of ELA/ELD standards. But this selection should also consider the student context. For one, knowledge acquired about emergent bilinguals and their families may help identify language practices that can be assets and areas for continued growth. You should know the students' *level* of English proficiency across the common modes of reading, writing, listening, and speaking by seeing results from state ELD tests. You can also use knowledge gathered using methods described in Chapter 4. The goal is to support all students in science-, language-, and literacy-rich and relevant learning opportunities. But the type and level of support will differ between students. All students should be expected to write an explanatory text. But you can expand the language, modes, and context in which students demonstrate their understanding. For instance, an emergent bilingual can write an evidence-based explanation as part of an informational website or presentation shared with the community. Knowledge about the assets that students bring could also inform support by framing learning around literacies more familiar and relevant to the specific emergent bilinguals in class. For example, some students might help their families construct online newsletters that need to reach and inform a broad audience. Given this asset, a culminating assessment might call on students to construct their own newsletters that translate the more complex evidence-based explanations into a more accessible form for others.

You may also learn of some emergent bilinguals who are in particular need of reading support, whereas other emergent bilinguals shine in this area. Other emergent bilinguals may be still emerging in their reading in English, but are quite advanced in reading in their home language. Finally, some emergent bilinguals may still be in need of reading support in both English and in their home language. This information would inform specific types of support (e.g., more extensive modeling, grouping, resources, use of translanguaging) that would differ among emergent bilinguals in the class. To be clear, the rich science, language, and literacy opportunities would be the same for all. It is the *type and level of support* that would differ.

CONCLUDING REMINDERS

- Weaving in ELA/ELD standards provides emergent bilinguals with opportunities to make meaning of rich science ideas while they are supported in interpretive, collaborative, and productive language.
- You can intentionally narrow the unit down to a few ELA/ELD standards that best align with the essential NGSS performance expectations.
- While ELD might focus on "English," you should also be aware of emergent bilinguals' primary or home language as a rich meaning-making tool.

CHAPTER 7

Anchoring the Unit With Phenomena, Texts, and Assessment

In this chapter we use Planning Tool 4: Framing the Unit Through Rich and Relevant Phenomena, Texts, and Assessment to move from abstract science ideas and language practices to the concrete and relevant world of naturally occurring events (i.e., phenomena) that can be interpreted and explained through activities and texts that are authentic to science. We also use Planning Tool 4 to plan a culminating assessment activity that gives emergent bilinguals the opportunity to demonstrate their science learning and language/literacy development in ways that apply and/or extend learning from phenomena and texts. For this chapter, consider the following guiding questions:

- How do you select phenomena and texts to anchor a unit?
- How do you provide emergent bilinguals with rich and relevant opportunities and language support while assessing their learning?

MOVING FROM STANDARDS AND BIG IDEAS TO PHENOMENA, TEXTS, AND ASSESSMENT

The big science idea for the Ecosystem Interactions and Resources unit is that *natural processes and human activity affect Earth's ecosystem.* This statement was generated by unpacking essential NGSS performance expectations and other relevant curricular resources. However, the big idea and related standards are still abstract. By *abstract*, we mean that they deal with concepts, principles, theories, and laws, not the concrete world as observed and experienced. The abstraction of big ideas and standards, and how science in general has often been communicated in the classroom, makes it less accessible and relevant to students. The same is true for language. The language practices of citing evidence, communicating orally with others, or writing an explanatory text are only relevant within a particular context. Cite evidence, communicate, and write about what? For what reason?

To begin making the big idea and standards richer in meaning and more relevant, you can first identify an anchor phenomenon. An *anchor phenomenon* is either (1) an observable and relatable event (e.g., a solar eclipse, sea level rising) or (2) a real-world problem (e.g., developing vaccines) that can be explained or solved with the help of scientific principles, theories, laws, and/or mechanisms. Students should be introduced to the anchor phenomenon at the start of the unit and revisit it repeatedly as they investigate, model, and eventually explain or propose a solution for the phenomenon. The importance for emergent bilinguals here is not just to make science more exciting, but also to make science relevant to their lives when their experiences have been ignored for too long.

Anchor phenomena also contextualize language. As students interpret texts related to the phenomena (e.g., a model or graph), collaborate with others (e.g., to hypothesize what is causing the phenomenon), and produce explanatory texts to account for what is happening, they need access and practice with discipline-specific language (i.e., vocabulary, syntax, and discourse) for sensemaking and for language/literacy development.

To better serve the synergistic role of supporting science learning and language/literacy development, we propose always accompanying an anchor phenomenon with an *anchor text*. We encourage you to use a journal or magazine article so that emergent bilinguals have the opportunity and support to interpret main ideas and cite evidence that can then be used, along with many other learning activities, to explore and explain the anchor phenomenon. This anchor text is ideally just one of many print, visual (e.g., graphs, diagrams), and digital texts that emergent bilinguals would interact with over the span of a science unit.

This purposeful use of science, language, and literacy through the context of an anchor phenomenon and text culminates in an assessment activity where students can successfully demonstrate what they have learned through a context or application relevant to their lives outside of school science. We also contextualize this culminating assessment by allowing students to produce and share texts that explain the anchor phenomenon using knowledge and linguistic competencies developed by interpreting the anchor text.

PLANNING TOOL 4: FRAMING THE UNIT THROUGH RICH AND RELEVANT PHENOMENA, TEXTS, AND ASSESSMENT

As shown in Figure 7.1, all three anchors—an anchor phenomenon, an anchor text, and a culminating assessment—are bound by the big idea already developed and framed through the curricular and the student context. Student learning throughout a unit would be anchored to this phenomenon, helping to craft a cohesive conceptual storyline, which also shifts the purpose of learning for students. Instead of learning for the sake of acquiring knowledge (including knowledge of how to do science), learning becomes about using scientific knowledge and practices to better understand the world around them.

Figure 7.1. Relating the Anchor Phenomenon, Anchor Text, and Culminating Assessment to the Big Idea

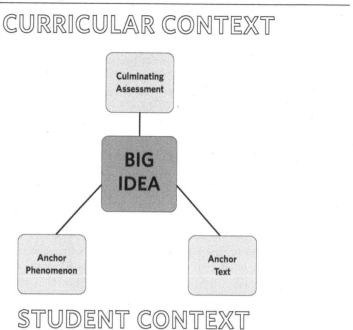

Planning Tool 4: Framing the Unit Through Rich and Relevant Phenomena, Texts, and Assessment (Figure 7.2), restates the big idea and then names or describes each anchor to showcase connections.

Anchor Phenomenon

When identifying an anchor phenomenon, focus on a phenomenon that is

- an observable event or real-world problem related to the big idea;
- rich in science (in need of a causal explanation or plausible solution); and
- relevant to students' lived experiences.

We anchored the Ecosystems Interactions and Resources unit around the observed phenomenon of how the abundance of sea otter, sea urchin, and kelp populations change over time in the California kelp forest ecosystem, represented as a graph obtained from the California Academy of Sciences (see Figure 7.3).

Figure 7.2. Planning Tool 4: Framing the Unit Through Rich and Relevant Phenomena, Texts, and Assessment

Guiding Questions
• What natural phenomenon or real-world problem would make the big idea rich and relevant?
• What texts could students be supported in interpreting throughout the unit?
• What culminating assessment activity could allow students to produce and share a text that explains the anchor phenomenon?

BIG IDEA:

Natural processes and human activity affect Earth's resources and ecosystem.

ANCHOR PHENOMENON	ANCHOR TEXT	CULMINATING ASSESSMENT
Observed abundance of sea otter, sea urchin, and kelp populations over time	"Ecosystem Superheroes: Sea Otters Help Keep Coastal Waters in Check" (*The Guardian*, adapted by Newsela staff, 11/14/2019)	*Ecologist Research, Blog, and Community Gathering*

Figure 7.3. Kelp Forest Population Dynamics Graph

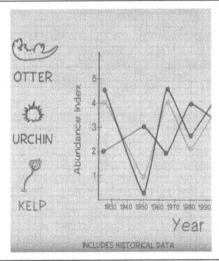

Note: From *Exploring Ecosystems: Coastal Food Web*, by California Academy of Sciences, n.d. (https://www.calacademy.org/educators/exploring-ecosystems-coastal-food-web).

The curricular context informs this phenomenon, which is rich in science. You can begin engaging students by showing them the graph and asking them to notice trends in the graph and to brainstorm the question, "What natural processes and human activity may be affecting this ecosystem and its resources?" Using knowledge learned in prior lessons and units, students can begin offering ideas for what natural processes and human resources impact these population trends (and why). Observing or experiencing a naturally occurring phenomenon, such as changing abundance of otters, urchins, and kelp, is metaphorically just the tip of the iceberg (as shown in Figure7.4). Underneath the surface lies a rich set of science concepts, principles, and theories that students would learn throughout the unit to explain the phenomena—what mechanisms, forces, and properties are accounting for the changes? In this way, the big idea connects the observable event to its underlying causal explanation.

There are numerous phenomena that would pique the interests of scientists, a science teacher, and students. The context of the kelp forest ecosystem was chosen intentionally because it came from the self-reported interests of students in the class (see Chapter 4). By knowing emergent bilinguals in the classroom, you can help students see connections between their interests and lived experiences and larger global and societal issues. For instance, scientists and local community members can use interaction and resource distribution patterns to predict, manage, and protect populations and resources in Earth's ecosystems. In Chapters 8–10, we will share examples of how this played out with students and how students engaged in science practices throughout the unit to make sense of and eventually explain the phenomenon.

Figure 7.4. Anchor Phenomenon Essential Question and Explanation

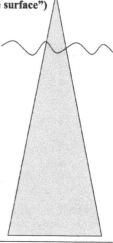

Anchor phenomenon essential question (what is observed "above surface")
"How can natural processes and human activity affect the kelp forest and its resources?"

Causal explanation or solution (e.g., what underlying science principles, theories, mechanisms help explain the phenomenon or solve the problem)
In an ecosystem, there are interactions between abiotic and biotic factors. All organisms are dependent on their environmental interactions with these factors. An organism interacts with its habitat and community to get the resources it needs to live, grow, and reproduce. Patterns in community interactions affect populations and resources in different ways. Organisms and populations with similar requirements for food, water, oxygen and other resources may compete with each other for limited resources, access to which consequently constrains their growth and reproduction. Uneven distributions of Earth's resources are also the result of past and current Earth processes.

Anchor Text

The second component is an anchor text. Even if you know where to search for texts (e.g., a particular website with resources or the library), you might not know exactly what text to choose. You want to use a text that

- relates to the anchor phenomenon and NGSS/Common Core standards.
- engages students and fosters curiosity.
- focuses on science or engineering practices, not just facts.
- provides students with ideas and evidence to help explain/propose a solution for the anchor phenomenon.
- is complex yet readable (with the appropriate literacy supports and limited to 1–3 pages of text).

Chapter 8 will introduce a planning tool to analyze the text chosen and decide how to use it for instruction.

We chose the Newsela article "Ecosystem Superheroes: Sea Otters Help Keep Coastal Waters in Check" as the anchor text for the Ecosystem Interactions and Resources unit. Newsela is an instructional content platform (see https://newsela.com/) that includes articles in English and in Spanish to supplement ELA, math, and social studies classrooms. Students access this article on Newsela, which has online tools that allow the teacher and students to annotate articles, read the article out loud in English and/or another language, and adjust the reading level for the same article.

"Ecosystem Superheroes: Sea Otters Help Keep Coastal Waters in Check" discusses research about the effects of a declining sea otter population, meeting the first guideline of relating to the anchor phenomenon. The context is likely to foster curiosity because both the anchor phenomenon and texts were chosen based on the students' self-reported interest in the coast. Moreover, the article itself describes the curiosity of a scientist asking the question: "Given its large appetite for urchins and shellfish, what happened to the rest of the ecosystem after sea otter populations declined?"

The article's purpose is to summarize research about sea otter population decline, including the history of that research. This purpose lends itself to focusing on science practices, not just facts. There are opportunities to identify a researchable question, methods with which to answer the question, and evidence to explain how natural processes and human activity affect kelp forests. The article directly addresses the varying factors affecting sea otter population growth (including human impacts) and the cascade effects these factors have had on this ecosystem.

Students can interpret and analyze this text by connecting it to their analysis of the anchor phenomenon graph. Since the article includes the year(s) that events occur in the kelp forest ecosystems, students can connect the text to patterns in the population graph over time. You can also support the Common Core ELA Standard "cite specific textual evidence to support analysis of science and technical texts" and the ELD standard "Reading closely and explaining interpretations and ideas from readings."

Consider the length and the complexity of the text. We recommend no more than three pages of written text (or about 1500 words) so as to focus more on deconstructing the text carefully, which takes time. "Ecosystem Superheroes" has 1325 words. The text itself is just over three pages, but that includes two visuals. Emergent bilinguals do not need to decode every word on the page. Instead, a text should be complex enough in science and language to provide opportunities for emergent bilinguals to practice, with support, making meaning of science ideas and deconstructing the text's structure, audience, and purpose. You can identify the text's lexile level to know if a text is at an appropriate reading level or not. You can find free lexile analyzers online (e.g., https://hub.lexile.com/analyzer). "Ecosystem Superheroes" has a lexile level of 1040, which puts it at a 7th-grade reading level. Then, by knowing your students' reading level, you can adjust language, such as vocabulary, as needed so as to not detract from the meaning.

Culminating Assessment Activity

You can culminate the unit with an assessment activity that ties to the anchor phenomenon and text, while allowing emergent bilinguals to apply the science and language practices they have learned throughout the unit. Although end-of-unit assessments might be associated with "exams" and "tests" that exclusively function to grade and report what has been learned (i.e., a summative use), we take the stance that learning and assessing are ongoing and always serve some formative role. What students and teachers learn through assessment, even if it happens at the end of a unit, should still inform next steps for learning in subsequent units. In particular, we suggest planning a culminating assessment activity that

- gathers and interprets evidence of NGSS and ELA/ELD performance expectations that extend ideas from the anchor phenomenon and text.
- allows emergent bilinguals to produce and share texts for a specific audience and purpose using a full linguistic repertoire.
- provides emergent bilinguals with appropriate language support and feedback.

Figure 7.5 summarizes the Ecologist Research, Blog, and Community Gathering assessment activity designed for the Ecosystem Interactions and Resources unit.

Figure 7.5. Ecologist Research, Blog, and Community Gathering Assessment Activity

Ecologist Research, Blog, and Community Gathering

- PART 1: Choose your ecosystem (location) and research information and data about your ecosystem using appropriate sources. (This should be based on family/student input with 6–8 local choices.)
- PART 2: As ecologists/community stewards, write Blog Posts to share with our community on the following topics:
  » Post #1:
    o A description of the **habitat/ecosystem** with a focus on **one species population** that lives there (include pictures and/or video)
    o What **resources** does this organism/population need to survive and thrive in this ecosystem?
  » Post #2:
    o Explain one or more important interactions/relationships this organism has in this ecosystem.
    o How might a natural process or human activity (pick one) affect this interaction? Predict how it might affect another ecosystem.
    o Predict how this interaction might affect other populations in this or other ecosystems.
  » Post #3:
    o How can you use what you have learned so far about how your organism interacts with other populations and resources in your ecosystem to design a solution to take better care of this organism and habitat?
- PART 3: Present and discuss ideas from your blog posts publicly during a community gathering with invited family/caregiver and community members.

Evaluative Criteria

- Appropriately describe ecosystem resources and interactions when making predictions about effects on ecosystems.
- Cite and use appropriate and various sources of evidence to support ideas.
- Produce written explanations structured to show cause-and-effect relationships.
- Ask questions and respond to questions from others to revise ideas.

Gather and Interpret Evidence of NGSS and ELA/ELD Performance Expectations

Pelligrino et al. (National Research Council, 2001) argue for the importance of aligning (1) the learning objective, (2) the instrument or activity used to observe student mastery of the goal, and (3) the evaluative criteria with one another. We next describe how to align all three assessment components for your own culminating assessment.

Culminating Learning Objective. A useful starting point is to integrate essential NGSS performance expectations and Common Core ELA and ELD standards to serve as the culminating learning objective. For the *Ecosystems*

Resources and Interactions unit, we decided that one NGSS PE in particular best served as the culminating objective: *Construct an explanation that predicts patterns of interactions among organisms across multiple ecosystems.* This NGSS PE incorporates the ELA and ELD standards of writing an explanation, which provide further guidance on what is expected when writing an explanation. We also wanted to make sure we assessed our interpretive language endpoint (citing evidence) and our collaborative language endpoint (interacting via written language).

Assessment Activity. You then want an assessment activity that will provide you with evidence of meeting the NGSS and ELA/ELD standards. Moreover, culminating assessments should not be thought of as stand-alone activities, but rather deeply embedded activities in the storyline generated through the anchor phenomenon and text. For the Ecosystem Interactions and Resources unit, the culminating assessment activity extended what students had learned about resources and interactions in the kelp forest into another ecosystem. Based on the outside locations and organisms that interested students and families, we created a research choice board where students could choose to research one Sonoma County local ecosystem type and one organism that lives there. Students researched and learned more about their ecosystem type and organism/population. Students moved away from class examples to evaluate how available resources affect populations and their interactions in their chosen ecosystem type.

What emergent bilinguals do in that assessment should also be a *meaningful* application or extension of what they have been working toward all unit long. In essence, it is not just assessing learning, but providing that culminating authentic experience that engages emergent bilinguals academically, metacognitively, and motivationally. For the Ecosystem Interactions and Resources unit, students extended their thinking to consider the impact of human activity and natural processes on these ecosystem resources and interactions. Their research culminated in explanatory writing pieces that they used to create blog posts to share and discuss as part of a community gathering with their colleagues, families, teachers, and community members.

The blog also provided an opportunity to leverage students' out-of-school lived experiences. As an example, one student shared how important hawks and hummingbirds were to his family and culture:

> In our native american tribe (coast miwok), males after they die are red tail hawks and females are hummingbirds. In the valley in between sonoma mountain and taylor mountain there is a lot of hawks everywhere.

One reason this is important is that students enter into writing through personal experience or background information research, but eventually fully explain the science.

Evaluative Criteria. You can then develop evaluative criteria that articulate what students would do to meet the culminating objective (see Figure 7.6). The criteria should align with the language performances that you are supporting, not just aligned to understanding science concepts. Because the written blog posts involve using evidence to explain conceptual ideas, we focused our criteria on three key areas, namely, how students (1) appropriately describe ecosystem resources and interactions when making predictions about effects on ecosystems, (2) cite and use appropriate and various sources of evidence to support ideas, and (3) produce written explanations structured to show cause-and-effect relationships.

You should also use rubrics to make expectations clear for students during the unit and to help students use feedback provided to them. When communicating evaluative criteria with emergent bilinguals, consider your own linguistic choices. Just like any other "text," the language should be accessible for students of varying reading proficiency. Avoid dense text, unfamiliar vocabulary, and bold key words or phrases.

Opportunities to Produce and Share Texts

Shepard (2013) posits that from a sociocultural perspective, assessment should do more than "[achievement] gap closing" and instead fully engage "the cognitive, meta-cognitive, and motivational aspects of learning" (p. xx). The blog was intended to help emergent bilinguals feel a sense of belonging within the classroom and school community; feel that who they are (including their language and lived experiences) are valued and matter; and have agency to create knowledge to inform their classmates and families. To foster that sense of belonging and agency, we hosted a community gathering (can be virtual or in-person). During the gathering, students connected with families and community members so that they could orally share main takeaways from their blogs and further develop ideas for how to act upon the proposed solutions. This gathering could also be in person, depending on what is more accessible for invited guests.

We encourage you to use multimodal culminating assessment activities. Through the blog post, students used a productive mode (i.e., writing the post), an interpretive mode (i.e., researching sources and reading classmates' posts), and a collaborative mode (i.e., interacting with others through peer review, comments, and the community gathering). All three modes aligned with the targeted ELA/ELD standards.

Encourage emergent bilinguals to use all available linguistic resources during the assessment activity. In producing blogs, students could incorporate words, pictures, diagrams, and links to videos. Emergent bilinguals could use any language to communicate these complex ideas, which might include writing some ideas in English and some in Spanish (or all ideas in both). This was particularly useful in the last stage of the assessment, where

Figure 7.6. Ecologist Research, Blog, and Community Gathering Evaluative Criteria

	Emerging Understanding and Language Use (i.e., not completely accurate and/or vague/generic)	Proficient Understanding and Language Use (i.e., appropriate but lacks elaboration and connections)	Advanced Understanding and Language Use (i.e., appropriate while also elaborating and making connections between ideas)
Conceptual understanding	Contains some inaccuracies when describing ecosystem interactions and resources. Rarely uses specific vocabulary (other than ecosystem, interaction, and resource).	Appropriately describes ecosystem interactions and resources. Uses appropriate and specific vocabulary (in addition to ecosystem, interaction, and resource).	Appropriately describes *patterns and relationships* in ecosystem interactions and resources. Uses and expands (e.g., defines, gives an example) on appropriate and specific vocabulary (in addition to ecosystem, interaction, and resource).
Use of evidence (interpretive language)	Mentions or cites any piece of evidence.	Cites two appropriate pieces of evidence.	Cites and connects two or more pieces of evidence to claims being made.
Structure and clarity of explanation (productive language)	Lacks (a) a meaningful claim or topic sentence and/or (b) sentences do not clearly show reasoning.	Includes (a) a meaningful claim or topic sentence and (b) sentences written and organized in a way that clearly shows cause-and-effect reasoning May have fragmented sentences, lack of transitions, or grammar/spelling/punctuation errors that make reading harder.	Includes (a) a meaningful claim or topic sentence and (b) sentences written and organized in a way that clearly shows cause-and-effect reasoning. Complete sentences, transitions, and none to few grammar/spelling/punctuation errors—making it easy to read.

community members (who themselves might be bilingual) would be able to engage in either language.

Provide Appropriate Language Supports and Feedback

You can provide appropriate language support to scaffold students' use of language during this ongoing assessment. We will detail some of these supports in Chapters 8–10. For instance, you can show example blogs and the particular purpose and audience across various blog posts. You can even contrast with other texts, such as lab reports, that might function to explain a phenomenon. You can provide graphic organizers and other tools to help emergent bilinguals as they learn to take research notes, evaluate resources, create a bibliography, craft blog posts, revise one another's work, and prepare to discuss during the symposium gathering.

As they move from research to draft blogs to final blogs and community sharing, you can provide students with feedback that pertains to the evaluative criteria and steps to learn from the feedback provided. Even if students are writing in a language other than English and you don't read that language, you can use tools such as Google Translate to understand and provide feedback in English or a language other than English. We also encourage you to provide opportunities for students to review one another's work and self-assess their work.

CONCLUDING REMINDERS

- Anchor the unit with a phenomenon, text, and assessment all aligned to the big science idea.
- Consider a phenomenon that is relevant to students and that is rich in science to explain what is happening.
- Consider a text that fosters curiosity, engages students in real science or engineering, and is complex yet readable.
- Use culminating assessment activities that (1) gather and interpret evidence of NGSS and ELA/ELD performance expectations while extending ideas from the anchor phenomenon and text, (2) allow emergent bilinguals to produce and share texts for a specific audience and purpose using a full linguistic repertoire, and (3) provide emergent bilinguals with appropriate language support and feedback.

PLANNING A CONCEPTUAL AND LINGUISTIC PROGRESSION OF LEARNING

Connecting a Conceptual Progression to Language and Literacy

Chapter 8 serves as a bridge between planning this whole-unit frame and planning a detailed sequence of learning activities that help emergent bilinguals progress in their conceptual understanding and language use. A key idea for this chapter is to start by mapping out a conceptual progression of learning. Then you can analyze the anchor text and identify additional language demands that you need to support emergent bilinguals in navigating during a unit. By connecting a conceptual progression to language demands you can arrive at a linguistic progression that pertains to interpretive, collaborative, and productive language. When reading this chapter, consider the following guiding questions:

- How do standards and anchors (i.e., phenomenon, text, assessment) help sequence conceptual ideas?
- Why is it important *not* to frontload vocabulary (i.e., not asking students to recall vocabulary before they make sense of conceptual ideas that pertain to the vocabulary)?
- How will it help you to analyze the anchor text as both a reader and an instructor?

FROM STANDARDS AND ANCHORS TO A CONCEPTUAL PROGRESSION

Our goal is to arrive at a clear sequence or conceptual checkpoints that deepen and progress student learning toward understanding the big science ideas and meeting the identified NGSS performance expectations. We do this through a backwards-mapping approach. Using *Planning Tool 2: Unpacking the Next Generation Science Standards and Curricular Resources,* we identified the culminating NGSS PE: *Construct an explanation that predicts patterns of interactions among organisms across multiple ecosystems.* We also arrived at the big science idea: *Natural processes and human activity affect Earth's resources and ecosystem.* From this, you can then draw on the key concepts found in the standards, textbook, and other curricular resources from Planning Tool 2 to sequence three to five conceptual "checkpoints" to culminate in explaining the big idea.

Instead of just being an exhaustive list of all concepts taught during a unit, these conceptual checkpoints force you to think carefully about the intentional *progression* of student learning. As seen in the first part of Planning Tool 5 (see Figure 8.1), we chose to write each checkpoint as a question that would eventually align with the assessments that would help check students' learning progress. For example, before exploring how biotic and abiotic factors *interact*, we wanted to make sure students knew about how ecosystems were *organized*. Many of the words used to describe concepts in any unit also relate to the language of crosscutting concepts, such as organization (structure and function) and affect (cause and effect). The use of crosscutting

Figure 8.1. Planning Tool 5: Conceptual and Linguistic Progression of Learning (Part 1)

Guiding Question

How can you generate a set of 3–5 checkpoints so that students are supported in progressing through the central science concepts and vocabulary en route to making sense of the big idea?

	Checkpoint #1	Checkpoint #2	Checkpoint #3	Endpoint
Concept	How are ecosystems organized? What resources are available in an ecosystem?	How do living matter (**biotic factors**) and nonliving matter (**abiotic factors**) interact in an ecosystem?	What relationships/ interactions exist in an ecosystem? How do these relationships/ interactions affect different populations and resources *within and across* ecosystems?	How can patterns in ecosystem relationships/ interactions be used to predict how organisms compete for and share ecosystem resources?
Relevant Vocabulary (be able to *use*, not just define, these)	Habitat, resource, ecosystem, population, community	biotic factor, abiotic factor, mineral, energy, groundwater/ freshwater, soil, natural (Earth) processes	relationship/ interaction (e.g., mutualism, predation, parasitism, commensalism), competition, analyze, interpret	human activity

concepts in particular helps keep the progression of learning grounded in the concepts (which involve relations and connections among ideas), rather than focusing on defining discrete vocabulary terms.

Next, we created a complementary progression for vocabulary terms. This vocabulary progression supports the underlying conceptual progression. Instead of teaching and expecting students to master all vocabulary terms *before* actually making sense of the conceptual ideas (also known as front-loading vocabulary), research supports the opposite approach. Students begin making sense of complex science ideas using a wide range of vocabulary (technical and nontechnical) along with other sensemaking resources that are more familiar to students. For instance, concepts such as "ecosystem resources" are rich in connections to students' prior knowledge and everyday, out-of-school lived experiences. Emergent bilinguals can use this more familiar language in English and in their first language to begin noticing resources observed through a video or in person, describing them, and making predictions or connections to other ideas. Then students can get targeted instruction that helps them develop the vocabulary and language that allows them to deepen their understanding and explain concepts more precisely and in ways that resemble the language the scientific community uses (Brown & Ryoo, 2008). This is not to say that you should *avoid* teaching vocabulary. Rather, consider what vocabulary a student really needs to know at this point and what they will be better able to use later in their learning. Vocabulary development will allow students to take the concepts they are making sense of with a variety of resources and then deepen their ability to make connections and communicate ideas.

PLANNING TOOL 6: ANALYZING THE ANCHOR TEXT

We next want to better understand the varied ways in which students would be using language (i.e., language demands) to both learn science and develop literacy. This connection between science and literacy best happens as emergent bilinguals interpret, discuss, and use the anchor text to further learning. However, in our own work *as* science educators and *with* science educators, we have found it challenging to actually use the anchor text to plan instruction that integrates targeted interpretive language support with science learning. In response, researchers and teachers in the Advancing Academic Achievement for English Learners (AAAEL) Project, funded by the Department of Education's National Professional Development program, developed the Preparatory Analysis of Text (PAT) planning tool (see Guilford et al., 2017), which we adapted into Planning Tool 6: Analyzing the Anchor Text for this book. You would complete Planning Tool 6 as you begin to move from unit- to lesson-level planning. Planning Tool 6 (see Figure 8.2) consists of the following three parts:

- *Part 1: Analyze the text as a reader.* While reading and rereading the text, mark words, phrases, and passages that are significant to eventually arrive at the most central point or points. These points should align with the unit's big idea. So the first part can be very useful in identifying whether this is an appropriate text or not.
- *Part 2: Analyze the text in preparation for instruction.* There are many conceptual and linguistic elements to the article that the science teacher might need to address and support students in unpacking and learning. These elements might include key concepts, required background knowledge or terminology, use of figurative language, purpose and audience, and genre and organization. Throughout, it is helpful to keep referring back to Planning Tools 2 and 3 to consider what standards could be addressed through this text.
- *Part 3: Develop leveled inquiry questions.* Each question would relate to one of the following three levels and guide student exploration and analyzing of the text themselves:
  - » Level 1: questions that ask for direct information from text
  - » Level 2: questions that ask the student to infer, interpret, and analyze based on text
  - » Level 3: questions that explore larger issues using and connecting outside knowledge and experience with learning from a text

Planning Tool 6 should also help you continue to recognize specific syntax and discourse (i.e., discipline-specific language demands) that students will need to interpret throughout the unit, not just in the anchor text, and eventually use during collaboration and when producing texts such as an explanation. In many ways, using Planning Tool 6 also models the approach to reading that you will eventually want students to use.

Figure 8.2. Planning Tool 6: Analyzing the Anchor Text

Guiding Question
How can we analyze the anchor text as a reader and as a teacher to better support interpretive language for students?
Title of Text: "Ecosystem Superheroes: Sea Otters Help Keep Coastal Waters in Check"
Author/Composer: *The Guardian*, adapted by Newsela staff on 11/14/2019
Source: *The Guardian*, adapted by Newsela staff
Topic of Lesson: Ecosystem Interactions and Resources

(continued)

Figure 8.2. (*continued*)

Part 1: Analyze the text *as a reader.*

Closely read the text, analyzing it as a reader by doing the following:

1.1. Mark significant words, notes, symbols, phrases, measures, and passages.

1.2. Identify the **THREE passages/quotations/sections** you believe will be the most significant for this lesson.

- "Islands with sea otters had healthy kelp forests. Islands without otters had barren sea floors littered with sea urchins and no kelp." (para. 14)
- "This constant activity masks a serious issue for the sea otter as an adult animal needs to consume vast amounts of food to survive. It needs to eat about a quarter of its own body weight—up to 24 pounds—every day."(para. 10)
- "The difference in annual absorption of atmospheric carbon from kelp photosynthesis between a world with and a world without sea otters is somewhere between 13 and 43 billion kg (13 and 43 teragrams) of carbon." (para. 17)

1.3. Identify the **THREE OR FOUR most important words/phrases** that will be the most significant for this lesson.

- *trophic cascade*
- *keystone species*
- *oscillating species numbers*

1.4. Add one more element of the text that resonates for you.

"Estes determined that commercial whaling after the Second World War was the cause. Before commercial whaling, killer whales fed on great whales of the North Pacific and southern Bering Sea, says Estes. By the time commercial whaling stopped, there were virtually no great whales left for killer whales to eat. So, they expanded their diet first to seals, sea lions and sea otters." (para. 23)

Human activity has impacted sea otter populations and the trophic cascade two unique times in history so far—hunting for pelts and whaling. How should humans interact in this (and other ecosystems) moving forward?

1.5. As a specialist in your discipline, describe what *you* take away from this text.

- All living and nonliving things in an ecosystem are deeply interconnected. In addition, interdependence exists between different ecosystems globally. When one ecosystem is changed or affected, it can impact other ecosystems globally. (Interdependence and interconnectedness are within *and* between ecosystems.)
- Balanced population sizes of one species in an ecosystem are dependent on resource availability and the other species in their community. Natural processes and human activity both play a key role in trophic cascades and can impact populations of species and resource availability in ecosystems.
- Humans have a responsibility to understand our roles, responsibilities, and impact on ecosystems and resources on Earth.

(*continued*)

Figure 8.2. (continued)

1.6. What would you want *your students* to take away from this text?

- A clearer understanding of
  » the connections and interactions that exist between all populations and abiotic factors in an ecosystem.
  » the relationship between natural processes, human activity, resource availability, and population size.
  » the role keystone species play in ecosystems.
  » the role and responsibilities of humans interacting in ecosystems.

Part 2: Analyze the text *in preparation for instruction*.

2.1. What **key concepts** in the text are *most important*?

- *Interactions* and *interdependence* in ecosystems
- *Trophic cascades* (impacts and how impacted in ecosystems)
- The importance and role of **keystone species** on ecosystems and overall global environmental health
- The *role of and responsibilities of humans* interacting in ecosystems

---

2.2. What **background knowledge** would students *need* before engaging in the text?

- What is a species? What does it mean to grow in numbers?
- What is a population? What is population size/species number?
- What is a cascade? What are "oscillating numbers"?
- What are predators and prey?

2.3. State whether there is one or more than one point of view, perspective, or voice in the text (from whose perspective?).

- There is one *main* voice and perspective (the author). At times, the text also describe Estes' perspective.

2.4. State any references to texts, ideas, and/or theories outside the text under study.

- Charles Darwin's *On the Origin of Species* is referenced at the beginning of the text.

2.5. State any **figurative language** used in the text.

- "this cascade of oscillating species numbers" (para. 2)
- "tightly interwoven components" (para. 24)

2.6. Describe the author's likely purpose for writing the text and intended audience.

- The author's purpose is to explain and evaluate the importance of keystone species, such as sea otters, in an ecosystem. Their purpose is also to judge and evaluate the impact of human activity on ecosystem health and balance.

Part 3: Develop leveled inquiry questions for the text.

Remember to ask students to use evidence from the text in their responses and the paragraph number(s) where evidence is found in the text.

Level 1: Questions that ask students for direct information from the text.

- Over 200 years ago, how large was the sea otter population in the Aleutian Islands? What caused that to change in the 1900s?
- What do sea otters eat? How much food do sea otters need to eat to survive?

(continued)

Figure 8.2. *(continued)*

Level 2: Questions that ask students to infer, interpret, and analyze based on the text.

- How are other populations in this ecosystem affected by sea otters? How are other ecosystems/areas around the world affected by sea otter populations?
- Why are sea otters considered a keystone species in this ecosystem?
- How have natural processes affected the trophic cascade, populations, and resources in this ecosystem?
- How has human activity affected the trophic cascade, populations, and resources in this ecosystem?
- Revisit the anchor phenomenon graph and deepen your response: Based on this graph and new evidence from this text, what natural processes and human activity do you think may be affecting this ecosystem and its resources? (Add potential causes to different points on the graph where the population size increases or decreases, using evidence from the text.)

Level 3: Questions that ask students to explore larger issues using/connecting outside knowledge and experience to ideas from the text.

- Based on your new learning from this text, how do you think humans should interact in this (and other ecosystems)? Why?
- How are keystone species and trophic cascades in one ecosystem important to the balance and health of other ecosystems on Earth?

PLANNING TOOL 7: UNPACKING THE UNIT'S SCIENCE-SPECIFIC LANGUAGE DEMANDS

Through previously used planning tools, you will have identified essential science vocabulary, language functions that complement science and engineering practices, and ELA/ELD standards that you can support emergent bilinguals in developing. Through an anchor phenomenon, text, and assessment, you have also encountered additional language demands that pertain to how scientists or engineers talk, write, and reason about, for a particular purpose and audience (i.e., discourse) and how scientists or engineers organize words or symbols to convey meaning (i.e., syntax). Now you can document these science-specific language demands in *Planning Tool 7: Unpacking the Unit's Science-Specific Language Demands* (see Figure 8.3). Planning Tool 7 allows you to then plan in a more intentional way by targeting and supporting the disciplinary language that emergent bilinguals will be navigating as they learn to interpret, collaborate, and produce text (i.e., the linguistic progression of learning).

Figure 8.3. Planning Tool 7: Unpacking the Unit's Science-Specific Language Demands

Guiding Question	
What science-specific language demands will be necessary during the unit for students to interpret, collaborate, and produce?	
Language Function (Active use of language for a particular purpose and audience, i.e., what you *do* with language)	• *Analyze* • *Interpret* • *Explain*
Discourse (How scientists or engineers talk, write, and reason about, for a particular purpose and audience, i.e., *tools* that convey *meaning*)	• *Representing data in a graph* • *Use of data analysis and textual evidence to account for a phenomenon*
Syntax (How we organize words or symbols in science to convey meaning (i.e., tools that organize)	• *Sentences that convey and link claim, evidence, and reasoning* • *Sentences that convey relationship and patterns found in data from a graph* • *Sentences that show cause and effect*

You have already begun to identify language demands as you unpacked curricular resources to arrive at key vocabulary (the basic ingredients for using discipline-specific language) to determine the primary ways students will use language *in action*—to serve some function in learning and doing science. Our focal science practices—"Analyze and Interpret Data" and "Construct Explanations"—also helped us identify ELA and ELD standards.

Next, you can take one of the language functions, such as "explain," and identify the related structure, tools, and/or norms that scientists use to talk, write, and reason about to convey meaning (i.e., discourse). For example, to "explain" the big idea, "how natural processes and human activity affect Earth's resources and ecosystem," students would need to develop a clear claim, list supporting evidence (and possibly counterevidence), and then coordinate the claim and evidence through reasoning with known science concepts and principles. This discourse (or way to make and communicate meaning) could be interpreted, used in collaboration with others, or produced—orally, in writing, or through some other modality. For this discourse, you can then identify the system or organizing words (i.e., syntax) needed to convey meaning while explaining. To coordinate claim-evidence-reasoning while writing an explanation, students would need to know how to construct sentences that convey and link (or transition between) claim, evidence, and reasoning. For example, students could see example phrases commonly used such as "evidence supporting this claim," "evidence suggests that," and "the reason for this evidence."

Figure 8.4. Planning Tool 5: Conceptual and Linguistic Progression of Learning (Part 2)

Guiding Question

How can you generate a set of 3–5 checkpoints so that students are supported in progressing through the interpretive, collaborative, and productive language on route to making sense of the big idea?

Mode of Communication	Checkpoint #1	Checkpoint #2	Checkpoint #3	Endpoint
Interpretive Language Progression	Identify key features of a population line graph.	Read closely to identify key ideas from a science article. Analyze a population line graph.	Interpret patterns and trends from a population line graph using content knowledge found in relevant sources.	Cite specific textual evidence to predict how organisms compete for and share ecosystem resources.
Collaborative Language Progression	Share something noticed or learned from a video or graph with a partner.	With a partner and available resources, orally discuss relationships or patterns (i.e., analyzing) found in a graph.	With a partner and available resources, orally generate interpretations of relationships or patterns found in a graph.	Orally and in writing, respond to comments and feedback about a shared explanation.
Productive Language Progression	Use familiar language to notice and communicate patterns in data and main ideas in a scientific article.	Write sentences that describe relationships or patterns (i.e., analyzing) found in a graph.	Write sentences that interpret a pattern or relationship found in a graph using content knowledge found in relevant sources. Use nouns and noun phrases to expand ideas and provide more detail about content knowledge.	Write a cohesive text that fully explains a phenomenon, supported with multiple related and relevant sources of evidence.

In describing these language demands, we are showing the connections among language function, discourse, syntax, and vocabulary that are needed to learn about and demonstrate the performance expectations. Multiple language functions will likely be expected of students during the unit, but, like the science practices, you want to focus in on one or two language functions that will be continuously targeted and supported. Again, the key is to be strategic in planning.

FROM ANALYSIS OF TEXT AND LANGUAGE DEMANDS TO A LINGUISTIC PROGRESSION OF LEARNING

The goal is to not only support emergent bilinguals in learning science, but also weave in rich and relevant support of their language and literacy development. Once you have identified the discipline-specific language demands for the unit, it will be easier to know exactly what you are supporting emergent bilinguals in doing so that there is a reciprocal relationship between science learning and the development of language and literacy. In the three chapters that follow, we take you through a progression of interpretive, collaborative, and productive language (see Figure 8.4) using sample learning activities and student work from the Ecosystem Interactions and Resources unit.

CONCLUDING REMINDERS

- Develop a conceptual progression through backwards planning—start with culminating NGSS performance expectations and use curricular resources to identify 3–5 conceptual checkpoints.
- Analyze the anchor text as a reader and in preparation for instruction to identify additional language demands that will help you organize a linguistic progression around interpretive, collaborative, and productive language.
- When analyzing the anchor text, also develop a set of leveled inquiry questions to scaffold science learning and literacy development toward deeper interpretation of texts.

CHAPTER 9

Interpretive Language Progression

In this chapter, we describe how you can support emergent bilinguals in making sense of texts (i.e., interpretive language) throughout a unit. We illustrate this with specific examples from learning activities planned for the Ecosystem Interactions and Resources unit. While reading through these examples, consider the following questions for your own units:

- What will emergent bilinguals be doing by the end of the unit to show how they can make sense of texts?
- How will interpretive language opportunities tie into the anchor phenomenon and text?
- How will you support and scaffold emergent bilinguals' interpretive language throughout the unit?

PLANNING AN INTERPRETIVE LANGUAGE PROGRESSION

The Common Core ELA standards provide culminating performance expectations that pertain to students making sense of texts (i.e., interpretive language). For example, in the Ecosystem Interactions and Resources unit, we identified *Cite specific textual evidence to support analysis of science and technical texts* (a Common Core ELA Reading in Science standard) as a focal interpretive language standard. You want to make sure language and literacy standards are woven into the culminating conceptual goal. The culminating conceptual goal for our unit was for students to explain *how patterns in ecosystem relationships/interactions can be used to predict how organisms compete for and share ecosystem resources.* Integrating the language and science goals led us to the following statement: *Cite specific textual evidence to predict how organisms compete for and share ecosystem resources.* We wanted students to be able to cite textual evidence from sources, such as articles and websites, while conducting research for the culminating blog-post assessment. Before being able to successfully cite textual evidence to support analysis of science and technical texts, emergent bilinguals need support in interpreting patterns and trends of graphs and charts they may find. And to interpret (i.e., make meaning) from patterns and trends, they need to both

Figure 9.1. Interpretive Language Progression

Checkpoint #1	Checkpoint #2	Checkpoint #3	Endpoint
Identify key features of a population line graph.	Read closely to identify key ideas from a science article. Analyze a population line graph.	Interpret patterns and trends from a population line graph using content knowledge found in relevant sources.	Cite specific textual evidence to predict how organisms compete for and share ecosystem resources.

know how to analyze graphs and charts (i.e., to identify patterns and trends) and read closely key ideas that will be used for interpretation. *Reading closely and explaining interpretations and ideas from readings* was another targeted interpretive language standard that came from the California ELA/ELD Framework. And before being able to analyze a population graph, emergent bilinguals need support in identifying the basic structure and function of a population graph.

We just described an intentional backwards-mapping process that wove in interpretive language and the conceptual focus (see Figure 9.1). Here we illustrate, sequentially, example learning activities from the Ecosystem Interactions and Resources unit that provide rich and relevant support for scaffolding emergent bilinguals' interpretive language so that you can apply the ideas to your own planned learning activities.

NOTICING AND WONDERING ABOUT THE ANCHOR PHENOMENON

To support interpretive language, you can first help students be aware of what they are learning and doing in the unit (i.e., exploring and explaining the anchor phenomenon), as well as introducing them to key texts that they will be interpreting. For the first learning activity of Ecosystem Interactions and Resources, we showed a short video clip of the kelp forest that depicted interactions among sea otters, sea urchins, and kelp. Videos provide emergent bilinguals with visual awareness of what they will be exploring before making sense of more abstract texts such as the population dynamics graph. When showing videos, we suggest turning on subtitles, if available, in English or in a language used by students other than English. We have also observed teachers providing emergent bilinguals with transcripts of the video in English and/or a language other than English. These subtitles and transcripts can be particularly helpful for those just beginning to learn English. Don't feel as though emergent bilinguals must decode every utterance from the video. They are beginning to make sense of the dynamic visuals and its eventual connection to later learning activities. Emergent bilinguals will differ in how much support is needed to make sense of the videos or other texts. Consider who needs the supports suggested in this chapter—you can always provide supports like sentence frames, but encourage emergent bilinguals with more advanced English proficiency to use their own wording.

You can then ask students to write down 2–3 notices and wonderings about the video, followed by an opportunity to share notices and wonderings in pairs or as a whole class. Notices and wonderings provide a low-stakes entry point into a complex text that cultivates curiosity and interests instead of focusing on right answers. During any call for notices and wonderings, encourage students to use whatever language or modality (i.e., pictures, diagrams) they want to describe what they notice, instead of looking for particular vocabulary words.

IDENTIFYING KEY FEATURES OF THE ANCHOR PHENOMENON

Now that students had seen a video on the kelp forest, we helped them identify features and the purpose of the population dynamics graph (see Figure 7.3). Graphs are a common text in math and science, with their own vocabulary, symbols, and discourse. Instead of looking for patterns in a sequence of words to make meaning from narrative text, students make meaning from a graph by looking for patterns and trends in the symbolic bars, lines, or scatter plots. Thus you play a role in helping students understand the discipline-specific language needed to interpret a graph.

You can start by asking students to list notices and wonderings for the graph. You can even model this process first by pointing out how you *notice* that there are three colored lines and *wonder* what each color represents (to keep it simple). Depending on students' prior knowledge of line graphs, you can also point out particular features of the graph on the document camera, such as the x axis or y axis and the movement of each line on the graph. You can also employ gestures. For example, you might ask students to move their hands in the air to represent the movement of the lines from left to right and ask for words to describe this movement. Depending on language proficiency, emergent bilinguals might be able to say "down, up, down, up" for the line representing the sea otter, or something more descriptive. Over time, you could help emergent bilinguals expand their vocabulary with words such as "fluctuate" or "alternate." You can also help them consider patterns by looking at two lines at the same time. As sea otters go "down, up, down, up," what is happening to the kelp? Students might realize that it also goes "down, up, down, up." Thus there is a relationship between the two.

We strongly encourage drawing on *cognates* to support this vocabulary development, meaning words that are similar between two languages. Cognates will become even more useful for emergent bilinguals with beginning proficiency. For instance, the Spanish word for *relationship* is *relación*. Frequent use of cognates helps value and in turn leverage emergent bilinguals' multilingualism. You will be helping them build confidence in that they can understand many science words because of their similarities with some other languages. Just be careful not to assume that emergent bilinguals know the science-specific vocabulary even when translated. This could become an opportunity to expand emergent bilinguals' multilingualism as they develop vocabulary in English *and* a language other than English. Overall, the explicit discussion around a graph's structure and purpose can elicit students' own language to help them develop vocabulary specific to analyzing graphs and awareness of how scientists organize data visually (syntax support). In the process, students also become aware of how information organized in a graph *conveys* meaning (discourse support).

You can elevate the initial discussion around notices through more targeted guiding questions. In our unit, we next asked students these questions:

- What patterns or trends do you *notice* in the population data for 1950 versus 1970?
- How do you think it relates to the *resources* available in that ecosystem?

To better understand what a *pattern* or *trend* means (e.g., something that repeats, occurs again and again), you can elicit patterns or trends your students have noticed or have heard about outside of school science. Students might provide examples such as tapestries woven as part of a family business, or TV shows that are "trending." This is an opportunity to connect with students' own experiences and language while helping them distinguish between different uses of these words. Eventually, you want to help students consider specific types of patterns and trends in a line graph—like direct or indirect relationships between variables.

WATCHING A VIDEO CLOSELY TO INTRODUCE KEY CONCEPTS

In a subsequent lesson, we wanted to use videos as the text to introduce the concept of ecological roles and relationships. The goal here is to learn conceptual ideas that students will need in order to eventually interpret the anchor phenomenon. But in the process, we are also weaving in language support to better know how to make meaning from a video.

We suggest using tools such as the graphic organizer depicted in Figure 9.2 for emergent bilinguals to organize ideas. If needed, you can provide more scaffolded graphic organizers depending on students' language proficiency. Our graphic organizer first asks students to describe roles (or niches) and relationships in their own lives to contextualize the important concepts of ecological niches and relationships. The graphic organizer also includes key definitions that students can draw on during a discussion to compare how the words are used in different contexts, thus expanding emergent bilinguals' nuanced understanding of these words. For example, a student may have a "role model" or play a "role" in a play—two uses of the word that are similar to, but not quite precise enough to fully describe ecological roles. Once again, you can introduce cognates. In Spanish, *role* is *rol*, whereas *niche* is *nicho*. We then played the video *Exploring Ecosystems: Coastal Food Webs*, produced by the California Academy of Science.

Like choosing articles, you should be intentional about the choice of videos. Instead of just serving to introduce students to the key organisms that will be studied (i.e., sea otters, sea urchins, kelp), *Exploring Ecosystems:*

Figure 9.2. Graphic Organizer for the Video *Exploring Ecosystems: Coastal Food Webs*

Before watching the video:

ROLE or **NICHE** - a *part, job, or function* an organism has in its ecosystem

ROL o **NICHO** - una *parte, trabajo o función* que tiene un organismo en su ecosistema

1. What are some roles or niches you have in your life? Explain. ¿Cuáles son algunos roles o nichos que tienes en tu vida? Explicar.

RELATIONSHIP - the ways in which **2 or more ideas, objects and/or organisms** are <u>connected and interacting</u>

RELACIÓN - las formas en que **2 o más ideas, objetos y/o organismos** están <u>conectados e interactuando</u>

2. Describe a relationship you have in your life. Explain. Describe una relación que tengas en tu vida. Explicar.

During the video:

3. Complete the Table Below.

ORGANISM	What **ROLE** does it have in this ecosystem? *(Why is it important to this ecosystem?)* ¿Qué ROL tiene en este ecosistema? *(¿Por qué es importante para este ecosistema?)*	What **RELATIONSHIPS** does it have in this ecosystem? *(How does it interact or connect with other organisms or abiotic factors?)* ¿Qué RELACIONES tiene en este ecosistema? *(¿Cómo interactúa o se conecta con otros organismos o factores abióticos?)*
Sea Otter Nutria de mar		
Sea Urchin Erizo de mar		
Kelp Quelpo		

4. What is a **PREDATOR**? What is a **PREY**? ¿Qué es un DEPREDADOR? ¿Qué es PREY?

5. What is being transferred in a food web?

6. What happens to sea urchin and kelp populations when sea otter populations are lower or absent? Explain.

Coastal Food Webs visually represented and described rich science ideas and the work of scientists pertaining to studying interactions (e.g., predator–prey) in the ecosystem. The video also embedded the population dynamics graph along with other ecosystem models. We asked students to list two "notices" and two "things learned." We showed the video again, now asking them to complete the rest of the graphic organizer depicted in Figure 9.2 to, among other prompts, capture roles and relationships observed in the video. Once again, the guiding questions are a type of language support to help emergent bilinguals know what to focus on and begin making sense of from the video. In subsequent lessons, students came to understand in more depth the type of interactions being observed in the video.

MAKE SENSE OF KEY VOCABULARY FROM A TEXT

While making sense of a video, article, or other text, emergent bilinguals might have a proficient understanding of previously learned terms (e.g., *population* and *resources* in our unit), but are likely less confident about new words introduced in the text (e.g., *trophic cascade* and *keystone species* in the video and anchor text). You can first use the strategy Raise Your Voice! (see Figure 9.3) so that students practice being able to orally say the word before they collaborate with others and produce texts with the vocabulary. In this strategy, you pronounce the terms in English and language(s) other than English spoken by emergent bilinguals in your class. You then ask the class to chorally repeat the terms—in both or all languages.

Figure 9.3. Raise Your Voice for Symbiotic Relationship (Relación Simbiótica)

This Week's Science Vocabulary

interaction *or* relationship (symbiotic relationship)
interacción o relación (relación simbiótica)

the ways that different species of organisms **effect, impact** or **connect with** each other in an ecosystem

+ = *positive impact* (helps or benefits the other organism)

- = *negative impact* (harms the other organism)

0 = *neutral* (no impact or other organisms is unaffected)

You can follow Raise Your Voice! with additional vocabulary support such as visuals, gestures, and analogies. For instance, you can represent the word *cascade* with gestures as hands are used to emulate a cascading waterfall and/or a picture of one. Cognates could be used as well (the Spanish word for *trophic* is *trofico/a*). The article students read for our unit mentions species being added or removed, which has a cascading effect through the various trophic levels of the ecosystem. Thus, making sense of the vocabulary serves the purpose of helping to make sense of the more central ideas.

As an example of how the science teacher might pause a video to clarify vocabulary such as *predator* and *prey*:

> *Teacher* [pausing video]: Let's talk about question four. What is a predator? Who's got a good idea?
> *Student:* It's something that hunts.
> *Teacher:* Something that hunts. So a predator is an organism that hunts or kills another organism in order to get its food.
> *Teacher:* Good. Everyone after me. Prey.
> *Students (chorally):* Prey
> *Teacher:* I want you to notice that in English there are two spellings for prey. So this [points to video] prey is p-r-E-y, not p-r-A-y (gestures something praying). So in the video, what's the prey?

In this example, the teacher starts by eliciting students' own idea (something that hunts), using it as an opportunity to clarify a homonym in English (*prey* vs. *pray*). The focus is not just to define *prey* but to better make sense of the video.

IDENTIFYING KEY IDEAS FROM THE ANCHOR TEXT

The next checkpoint in our interpretive language progression was to "Read closely to identify key ideas from a science article" and to "Analyze a population graph." As a reminder, we chose the Newsela article "Ecosystem Superheroes: Sea Otters Help Keep Coastal Waters in Check" as the anchor text because it

- engaged students and fostered curiosity.
- focused on science or engineering practices, not just facts.
- provided students with ideas and evidence to help explain/propose a solution for the anchor phenomenon.
- was complex yet readable.

Students in our class were already familiar with the text annotation norms shown in Figure 9.4. We also asked students to read the article once through

Figure 9.4. Annotation Guide

Annotation	Meaning/Purpose
Circle	Powerful words and phrases
Triangle	Content vocabulary
Underline/Squiggly underline	Confusing words, ideas
Writing in the margins	Recording of questions particular parts of the text raise and perhaps even the answers
Symbols in margins	Sequence of points being made

on their own for homework. The purpose of this first read was not to get any deep meaning from it. It was just to become familiar and perhaps for students to realize how much they don't know from a first read. You can alternatively use a prior class for students to begin reading the article. And you may need to provide more explicit modeling and guidance about how to annotate using norms such as the ones we provided. We recommend re-establishing these annotation norms anytime an article is read and continue to model using the norms for students.

For us, the annotation always began by numbering each paragraph (e.g., 1, 2, 3). You can help the whole class number paragraphs the same way by stating "Paragraph #1" and using a document camera or other visual tool to point to the first word of the paragraph. You can then ask students to chorally repeat the first word (e.g., *ecosystem, trophic, Estes, the island, the sea otters, sea otters*). This simple exercise provides the class with an opportunity to start noticing what the article is about, evidenced by how *sea otters* and *Estes* (the scientist) are the first word of multiple paragraphs. You can also remind students of the purpose of numbering paragraphs—so that the class has a reference point for talking about ideas in the article and can cite the paragraph number to indicate where they found textual evidence to support responses to questions posed.

Next, you can model annotations of the first half of the article to the whole class. This modeling supports meta-awareness of what readers do while reading closely, or as told to students, "I'm going to underline words and phrases that my brain thinks are important." You can also encourage students to add these annotations to their own initial annotations. As an example of how this might sound in a classroom:

Teacher: [Reading aloud] *James Estes is an American marine biologist* [underlines biologist]. *He has studied wildlife in the North Pacific Ocean for the past 45 years* [underlines 45]. [Commenting on the sentence]. That is a long time to focus on one area. In my mind that makes me think about how much of an expert he is. [Reading] During that time, he has shown how predators can *change* their environments [underlines sentence and circles predator]. [Commenting on the sentence] And that feels important to me, how predators can change their environment. [To students] What happens when you see words repeated and circled? What does your brain do?

Student: They stand out.

Teacher: They stand out. My brain starts to see a pattern.

This type of modeling is more than just helping students follow procedures. The modeling is giving students a window into the mind of a reader and helping them begin the journey of making sense (or interpreting) what is being read. They will be learning some vocabulary, but mostly how the article is structured in a particular way to convey meaning (or discourse).

USING LEVELED INQUIRY QUESTIONS TO ANALYZE PATTERNS AND TRENDS

After modeling how to annotate half of the article, you can provide students with the leveled inquiry questions that you generated using Planning Tool 6: Analyzing the Anchor Text. We have provided a page from the student handbook that included these questions in Figure 9.5. We suggest establishing and reviewing norms for when students write responses to leveled questions, such as using complete sentences with a paragraph number in parentheses to note where the information was found. The sequence of questions will scaffold emergent bilinguals in accessing and then getting deeper and deeper meaning from the text. The questions will also scaffold emergent bilinguals toward being able to use evidence from the text to analyze and interpret the kelp forest population graph.

The two Level 1 questions are intended to be found directly in the text. For additional support, you can also ask students how *they* would begin the sentence. Students might recognize that they can look at the question itself and begin by repeating the question "The sea otter population was . . ." To model this process, you can ask a student to read the question and then ask everyone to point with their finger or pencil to the paragraph where they can find the best answer. In Chapter 10, we describe how to use a pair-share grouping structure so that student collaboration supports making sense of texts. The responses to Level 1 inquiry questions can also serve as a check for understanding at the end of a lesson (or what might be known as an exit ticket). We will then return to these leveled inquiry questions in Chapter 11 to discuss productive language use.

Figure 9.5. Leveled Inquiry Questions

Level 1: questions ask for direct information from text

- Over 200 years ago, how large was the sea otter population in the Aleutian Islands? What caused that to change in the 1900s?
- **WHAT** do sea otters eat? **HOW MUCH** food do sea otters need to eat to survive?

Remember to use <u>evidence from text</u> in your responses. Remember to include the paragraph number (s)where evidence is found in the text.

Level 2: questions ask the student to infer, interpret, and analyze based on text

- How are other populations in this ecosystem affected by sea otters? How are other ecosystems/areas around the world affected by sea otter populations?
- Why are sea otters considered a *keystone species* in this ecosystem?
- How have *natural processes* affected the *trophic cascade*, populations and resources in this ecosystem?
- How has *human activity* affected the *trophic cascade*, populations, and resources in this ecosystem?
- Revisit the Anchoring Phenomenon graph and deepen your response:
  » Based on this graph and new evidence from this text, what **natural processes** and **human activity** do you think may be affecting this **ecosystem** and its **resources**? (Add potential causes to different points on the graph where the population size increases or decreases using evidence from the text.)

Remember to use <u>evidence from text</u> in your responses. Remember to include the paragraph number(s) where evidence is found in the text.

Level 3: questions explore larger issues using outside knowledge and experience

- **Based on your new learning from this text, how do you think humans should interact in this (and other ecosystems)? Why?**
- How are keystone species and trophic cascades in one ecosystem important to the balance and health of other ecosystems on Earth?

Remember to use <u>evidence from text</u> in your responses. Remember to include the paragraph number(s) where evidence is found in the text.

In a later activity, we intentionally wove together the *Exploring Ecosystems: Coastal Food Webs* video, kelp forest population dynamics graph, and "Ecosystem Superheroes" article. We showed more of the video, which displayed the same kelp forest population dynamics graph and invited students to predict relationships. However, the graph continues the trend lines beyond what students had previously seen (i.e., new data). At this point we stopped the video again to help students interpret "What's happening" in the graph. We reviewed basic features of the graph ("What does it mean when the line goes down?") and helped students notice the new data:

Teacher: Now we have a *new* pattern. What does it mean when the line
    goes straight?"
Multiple students: It's the same
Teacher: Same, or stable. Now, time to answer the question: What
    would happen if sea otter populations went down or were absent?
[Students write individually in their notebooks]

You can remind students of the wonders they already had recorded: "What's interesting here is that you all had some interesting questions about what might happen after 1990 and now this graph shows some interesting trends." We pulled on key ideas from the "Ecosystem Superheroes" article to help make these predictions. This sets them up to consider what might be affecting this ecosystem and the organisms living there. In later activities, students would conduct investigations and engage in more learning activities to deepen their conceptual understanding. In turn, this deeper conceptual understanding allows them to go deeper and deeper in being able to make sense of the texts they are reading.

CITING EVIDENCE FOR THE CULMINATING ASSESSMENT ACTIVITY

In the Ecosystem Interactions and Resources unit, students progress from identifying key features of texts to analyzing and interpreting information (main ideas and data) to finally locating and citing textual evidence in support of producing and sharing blog posts. To accomplish this culminating task, students would apply what they learned about reading texts closely and analyzing articles and graphs to help them research an organism's habitat, available resources, and interactions with other organisms.

You can support emergent bilinguals in conducting their own research like this by modeling how first to find and evaluate articles. We used the template shown in Figure 9.6 as a checklist for evaluating sources. Then we suggest preselecting three contrasting articles to help students see the differences between less and more credible sources. We also suggest modeling and requiring students to find one source that includes a chart or graph to analyze and interpret. This way, students apply what they learned from working with the kelp forest population graph.

You might need to help students find articles appropriate for their reading level and revisit what kind of charts or graphs they might encounter. For example, you could give emergent bilinguals with lower English reading levels articles at an appropriate lexile level, but remember to make it a complex enough text where they can practice inferring and finding evidence. You can also provide articles in languages other than English. Students with

Figure 9.6. Evaluating Sources Handout

Evaluating Sources: Is this a Credible Source of Information?

1. What is the <u>purpose or motivation</u> of the website?

   - Is the main purpose of the website to inform or is it to sell/persuade? (*Review the* <u>content</u> *and the* <u>writing style</u> *of the website or source.*)
   - *Be sure the websites you use for information are to inform (NOT sell/persuade).*

2. Is the content backed up by <u>proof or evidence</u>?

   - <u>How</u> was the information on the website gathered?
   - Are citations provided that indicate the information was collected from credible sources (such as <u>peer-reviewed journal articles</u>)?

3. Is the content coming from <u>a credible source</u>?
   (long-standing, well-respected, trustworthy, or prestigious source)

   - <u>Look at the URL of the site.</u> **.gov** and **.edu** websites tend to be reliable and thus credible.
     - *Be careful with commercial websites (with the **.com** domain). The information they provide may not be false, but it might be biased or incomplete.*
   - Try Google Scholar, Science.gov
     - May be more advanced, but reliable

4. Is there evidence that the content is written by experts?

   - Look for information about the <u>author's' credentials.</u>
     - Are they an expert in the topic? What degrees/experience do they have (BA, BS, MA, PhD, etc.)?
   - If there is no author, use evidence you've collected about the website.
     - Is it a government website? University?

---

similar ecosystems could also share articles found with each other and give feedback on each other's evaluation of credibility.

You can also support students in using digital or hard copy "notecards" to capture key ideas, quotations, and evidence from credible resources. Students could additionally use notecards to analyze and interpret a chart or a graph. Notecards help students put ideas into their own words and images, which they can then use and refine for their actual blogs. While locating and citing evidence for their blog posts, emergent bilinguals have the opportunity to more deeply understand the big science ideas. At the same time, the anchor phenomenon, anchor text, and culminating assessment provide a context for better being able to make meaning (not just decode words from) texts.

CONCLUDING INTERPRETIVE LANGUAGE REMINDERS

- **Position Students as Sensemakers:** Provide opportunities and support for emergent bilinguals to make sense of a wide range of discipline-specific texts that engage them in science practices such as using models, analyzing and interpreting data, and evaluating information.

- **Facilitate Productive Student Discourse:** Help emergent bilinguals understand how texts are structured to convey meaning. For instance, how does the structure of a line graph convey patterns and relationships? Use talk between students (e.g., pair-shares) to better help emergent bilinguals read texts.
- **Scaffold Language and Literacy:** Use leveled inquiry questions to help emergent bilinguals progress from just noticing or identifying features of a text to understanding main ideas to applying ideas to the anchor phenomenon. Support emergent bilinguals' development of vocabulary, syntax, and discourse through a variety of strategies such as cognates, sentence frames, and modeling.
- **Contextualize Science Activity:** Base the anchor text on known student interests or out-of-school, lived experiences. Keep coming back to the anchor phenomenon.

CHAPTER 10

Collaborative Language Progression

You are likely to expect students to work together throughout a science unit to, for example, conduct investigations, share ideas, and give one another feedback. In this chapter, we describe how you can help emergent bilinguals progress in the language needed to make meaning *with others* (i.e., collaborative language). Once again, we illustrate this with specific examples from learning activities planned for the *Ecosystem Resources and Interactions* unit. While reading through these examples, consider the following questions for your own units:

- What will emergent bilinguals be doing by the end of the unit to show how they can make meaning with others?
- How will collaborative language opportunities tie into the anchor phenomenon and text?
- How will you support and scaffold emergent bilinguals' collaborative language throughout the unit?

PLANNING A COLLABORATIVE LANGUAGE PROGRESSION

In many ways, collaborative language naturally builds on interpretive language. We always try to give emergent bilinguals individual time to read or listen, think, and even write before asking them to engage in back-and-forth verbal exchanges with their peers. For emergent bilinguals, collaborating with other students presents a new set of challenges, namely in that they need to rapidly interpret what others are saying and then respond. But with that challenge comes new opportunities to hear other ideas and how others use language, and engage in continuous feedback as emergent bilinguals make deeper meaning of ideas. And that is really the goal—to have the tools and practice in exchanging ideas with others, since there is rich meaning-making that happens with others.

One challenge for planning a collaborative language progression is that those expectations do not show up easily in Common Core Literacy in Science and Technical Subject standards. In part due to the shift toward distance learning during the COVID pandemic, we targeted *written* collaboration

Figure 10.1. Collaborative Language Progression

Checkpoint #1	Checkpoint #2	Checkpoint #3	Endpoint
Orally and in writing, share something noticed or learned from a video or graph.	Orally and in writing, discuss with a small group relationships or patterns (i.e., analyzing) found in a graph.	Generate interpretations of relationships or patterns found in a graph with a partner.	Orally and in writing, exchange ideas about explanations produced, read, and heard.

that can supplement and support emergent bilinguals in verbal exchanges. We looked toward the California ELA/ELD Framework to identify our focal collaborative language standard—*Interacting via written language (print and multimedia)*. We assessed students' collaborative language as they exchanged ideas orally in the Ecologist community gathering and via written comments from their blog posts. To develop the discipline-specific language needed to collaborate, they need opportunities to interact with pairs and with small groups for the purpose of generating new ideas and interpretations together (i.e., collaborative inquiry). And to engage in collaborative inquiry, they needed opportunities to share what they had noticed and learned from various texts. Once again, this is an intentional sequence of checkpoints (see Figure 10.1) that will involve continuous listening to and assessing emergent bilinguals.

PAIR-SHARES AND INTENTIONAL GROUPING

Sharing ideas with another student (i.e., a pair-share) is a useful collaboration entry point for emergent bilinguals. While planning partner assignments, be mindful of who the emergent bilinguals in the class are and the particular assets they bring. It is important for emergent bilinguals to eventually interact with students of varying linguistic assets. When first learning to make meaning from a science article, you might purposefully partner emergent bilinguals who are more fluent readers in both English and Spanish with those who are still working on their reading in both languages. Beyond purposeful grouping, you can provide explicit norms and guidance for how partners interact with each other. For example, after students recorded what they noticed and learned from a video, we assigned each partner a letter ("A" or "B"). We asked "A"s to share one thing they noticed or learned, and then the "B"s shared. There is little expectation for back-and-forth dialogue here. Emergent bilinguals are just getting comfortable interacting with others using any language available.

TIME TO THINK AND WRITE

Before asking emergent bilinguals to share complex thinking and ideas orally, we encourage you to give them individual quiet time to think and write. This process time allows them to articulate thinking before being asked on the spot to produce language orally. For individual writing, you can provide sentence frames that scaffold writing sentences that convey patterns and wonderings about what happened. For example, we employed sentence frames to help all of our students first individually analyze the kelp forest population dynamics graph. Examples include . . .

- *Between _____ year and _____, I noticed_____.*
- *The _____ increases/decreases when _____.*
- *One relationship I notice in this data is between _____.*
- *I wonder why/what happened when _____.*

Then, when placed with a partner or with small groups, we reminded our emergent bilinguals in particular to refer back to these responses so that they can be shared with the group. You can encourage emergent bilinguals with more advanced proficiency to brainstorm their own sentence frames. In later lessons, these sentence frames may be available to all students, but they would eventually be encouraged (and expected) to write without the support. Sentence frames are always meant as a temporary support that can eventually be removed or modified to introduce more sophisticated and varied phrases. We would also want them to expand on the sentence written and shared with others.

COLLABORATION BOARDS

You can use collaboration boards to support written exchanges among students. A collaboration board first asks all students to post a written response to a question. The collaboration board can be physical (e.g., Post-it notes on a whiteboard) or digital (e.g., Nearpod, Jamboard). You would then ask students to read the collection of responses and ask them to comment on one or a few of them. The purpose of the collaboration board is again to provide emergent bilinguals with time to think and process ideas that they are sharing on the collaboration board without first having to engage in more intensive interpretation of oral language as they listen to multiple peers and the teacher. In essence, the collaboration board primes emergent bilinguals to draw on a range of linguistic resources so that they can more easily navigate the quicker back-and-forth collaboration typically encountered in small-group or whole-class discussions. We were also intentional about using collaboration boards to support our unit's targeted collaborative language standards—*to interact via written language (print and multimedia)*.

You can encourage emergent bilinguals to use any language available to them and use pictures or diagrams on a collaboration board. You also want to ensure that they respond to other posts. For example, in our unit, we asked a question from the anchor text, "According to the reading, what other populations are affected by sea otters and how?" One emergent bilingual wrote, "Sea urchins are affected by sea otters because sea otters consume sea urchins in order to survive." Another student commented with "That's interesting, what paragraph did you find that on?" You can acknowledge in the whole-class setting how this comment is encouraging us to consider the source of information. You can ask for patterns and trends in responses and use this to assess any gaps in knowledge. By writing posts, reading one another's posts, and responding to them, students are making meaning with one another (collaborative language) even before they start talking with one another.

SCIENCE TALKS FOR COLLABORATIVE THINKING AND SHARING

You can continue to support emergent bilinguals' collaborative language development through whole-class or small-group *science talks*. Science talks are more intentional than just impromptu questions and discussions. Productive science talks include purposeful, planned, open-ended questions, aligned with the learning objective. Science talks also include planned structures, norms, and resources for students to exchange ideas with one another. Our science talks always start with visual and oral reminders (Figure 10.2) about norms/expectations and science talk moves. The science talk moves are designed to help students focus on how to structure ideas (syntax) to reason and convey ideas that connect with the specific language function. For instance, we include sample talk moves such as "How do you know that? What is your evidence for that?" that support using language to analyze, which will differ from other learning activities that focus on predicting.

You can listen to each group's conversation and use preplanned (or back pocket) questions to further probe students' thinking. Your own talk moves and back pocket questions should also align with the focal language function when interacting with small groups. At the beginning of a unit, we encourage talk moves and back pocket questions that elicit students' everyday experiences and language they would use to describe what is happening—instead of expecting precise vocabulary. You can also employ gestures and more visuals depending on emergent bilinguals' proficiency in interpreting oral language. The most important part, especially for emergent bilinguals just entering the classroom, is to feel valued and part of the community—even if many of their peers and possible their teacher only speak English.

Figure 10.2. Science Talk Norms and Moves

**All student ideas are *respected, important* and *"fair game"* for examining and discussing (even if ideas are not correct or incomplete). *As a community, we all help each other learn and think more about our ideas!*

**Science Talk is an important way scientists and engineers learn, develop new ideas, and deepen thinking on a topic.

Science Talk Moves to Try On:
- Can you say more about that? Can you share more what you are thinking? *(ELABORATE)*
- How do you know that? What is your <u>evidence</u> for that?
- I agree with _____ because _____.
- I disagree with _____ because _____.

Norms, expectations, and appropriate talk moves help emergent bilinguals be active participants. To further ensure that emergent bilinguals' thinking is visible to others, we encourage collaboration through students' full linguistic repertoire. This means that you encourage emergent bilinguals to use any language and encourage them to communicate with additional modes (e.g., oral *and* writing) or modalities (e.g., with a visual or use of technology like Google Slides). Using Google Slides as a collaborative space, emergent bilinguals could orally communicate ideas with a group recorder chosen to write on the Google Slide. They can exchange ideas in modes other than oral language. For example, emergent bilinguals could write directly on a Google Slide. In a distance learning environment, they could write in a chat window for the recorder to write on the slide.

FROM GROUP WORK TO COLLABORATIVE INQUIRY

Instead of just asking students to work together to answer questions that can be found directly in an article or the textbook, we emphasize *collaborative inquiry*. By *inquiry*, we mean that students have creativity and choice to generate new ideas, interpretations, or solutions together. Leveled inquiry questions help scaffold students toward interpretation and application of ideas. Level 2 questions require some inference or indirect evidence from the text. Because they are more open for interpretation, it is even more beneficial for students to collaborate and exchange ideas.

We suggest a clear guiding question, product, resources, and norms/expectations during any collaborative inquiry. To help make deeper sense of the "Ecosystem Superheroes" article students read, we posed the following guiding question that aligned with our big idea: *What natural processes and human activity do you think may be affecting this ecosystem and the organisms living there?*

To reinforce key vocabulary that students should use, you can make associations with pictures or objects. For instance, we consistently held up a globe of the Earth to signal "natural process" and a picture of a car to signal "human activity." You can also elicit students' prior ideas orally: "What are some natural processes we learned about? Now what about human activities?" Students would revisit examples already written down (e.g., changes in population sizes, pollution, bacteria/sickness, hunting). Our students previously used an online discussion board to brainstorm potential human activities that could influence ecosystems. We projected the discussion board responses and read them aloud, such as "a lot of people were killing the animals" and "hurricanes were blowing some dust and perhaps some pollution and that pollution might be causing this."

Now, with others, emergent bilinguals can then start connecting general ideas (in this case about human activity) to the anchor phenomenon. In Figure 10.3, we share actual notices produced from groups with emergent bilinguals about the kelp forest population dynamics graph. Like collaboration boards, this sharing provides you with important opportunities to assess students' science learning and use of discipline-specific language. For instance, you could probe or clarify the language used to indicate a pattern, such as "went up" or "go hand in hand." You might take this opportunity to point out to others how this is figurative language and introduce new vocabulary to communicate patterns more precisely. The class might even get to phrases, such as "there is a relationship between the kelp and sea otter populations," which can lead to questions such as "is there a positive or negative relationship?"

Figure 10.3. Group Notices and Wonderings

What patterns do you NOTICE in the population data for 1950 versus 1970? How do you think it relates to the RESOURCES available in that ecosystem?

Round 1 Collaborative Discussion Themes (NOTICES/WONDERINGS)

- Between 1950 and 1970, the sea otter population went up. (1950—sea otter population dropped to its lowest amount, 1970—sea otter population was at its highest)
- Kelp and sea otter populations go hand in hand.
- When the sea urchin population goes up, the otter and kelp populations go down. When sea otters stop eating sea urchins (start eating other stuff), it affects the other populations.
- When the sea otters dropped, the kelp dropped, and when the otters grew, kelp grew. When sea otters dropped, sea urchins grew, and when otters grew, sea urchins dropped.
- The otter population went down 1950 and went up 1970 meaning that they were endangered and people had to stop killing so they can breed and the population went up because no one killed them.

In exchanges such as this, you are helping students develop discipline-specific language by leveraging the language students used to describe patterns. Furthermore, students are expanding their linguistic repertoire by better understanding figurative language that might appear in other contexts.

In subsequent lessons, the meaning-making that happens with others can help emergent bilinguals produce discipline-specific texts (individually or collaboratively), which we will discuss in Chapter 11. The collaboration can continue as students share and discuss texts that they produced. For this collaboration, we facilitated a group "jigsaw." In one jigsaw structure, a student from each group is assigned as the "spokesperson" and stays at the group's whiteboard (or poster), as shown in Figure 10.4, while the remaining group members travel to the other groups and listen to the respective spokespersons. Then each home group rejoins and exchanges what they learned from others. This form of distributed expertise allows for sustained and widespread talk and interaction with other students—opening up even more learning opportunities for emergent bilinguals. Additionally, you can circulate to clarify any points and provide feedback on the spot—both in conceptual understanding, but also on vocabulary use, syntax, or discourse. For instance, you can notice the phrases and words emergent bilinguals use to signal "causes and effects" (i.e., a common discourse of science).

Figure 10.4. Sharing a Collaborative Interpretation

TARGETING COLLABORATIVE LANGUAGE IN THE
CULMINATING ASSESSMENT ACTIVITY

Culminating, end-of-unit assessments might be thought about as purely individual activities, where students show what they know and can do on their own. While you want assessment of individual thinking and performance, you should also not forget that collaborative language is an expectation that needs to be assessed throughout and at the end of a unit. Since our focal collaborative language expectation was for students to interact with written language (print and multimedia), we asked students to collaborate during the Ecologist Research, Blog, and Community Gathering through comments on the blog posts. The comments provided a mode, other than oral, to give feedback, make connections to student experiences, and deepen learning.

Before writing the blog, we gave a minilesson on writing quality comments on other students' blog posts. One key was to help students understand the purpose of comments: to interact with a classmate as a way to collaboratively make meaning (supporting discourse). Comments could add new knowledge, make a personal connection, and/or ask a question to deepen learning. Sentence frames can guide students in the discourse of writing comments that meet these purposes. For example, "I read your blog and it made me think . . ." to convey new knowledge. Without targeting a particular purpose, like adding new knowledge, students may fall back on generic comments, such as "good job!" or "interesting," which do not support meaning-making or provide further opportunities for language development. Students also learn how responding back to a post can deepen learning through more dialogue and perspectives.

Finally, our assessment culminated with an in-person or virtual community gathering where families and community members could dialogue with students about their proposed solutions to see if the class could be agents of change. You can not only have students rehearse how to present, but also develop questions they may have to get input/feedback from those attending on their ideas. Even if not given a grade, you can still listen in and provide feedback on how students were able to ask questions and respond to questions and ideas to continue supporting collaborative language use.

CONCLUDING COLLABORATIVE LANGUAGE REMINDERS

- **Position Students as Sensemakers:** Collaborative inquiry allows emergent bilinguals to have creativity and choice when generating new ideas and interpretations with others.
- **Facilitate Productive Student Discourse:** Use science talk structures and leveled inquiry questions to facilitate discussion between students that scaffolds toward deeper thinking and application.

Consider additional modes of collaboration beyond oral, such as collaboration boards.

- **Scaffold Language and Literacy:** Throughout a unit, provide richer opportunities for collaboration with appropriate support. Emergent bilinguals might start by just sharing a notice with a partner, and at the end by being able to generate interpretations with others. Provide sentence frames, graphic organizers, or other supports, but work toward helping students need these supports less and less over time.

- **Contextualize Science Activity:** Make collaborative inquiry more relevant. You can plan for emergent bilinguals to generate new ideas and interpretations with other students about a phenomenon or problem of interest. Better yet, elicit from emergent bilinguals what they want to explore, explain, and solve.

CHAPTER 11

Productive Language Progression

In this chapter, we describe how you can support emergent bilinguals in using what was learned by making meaning of texts and with others to now progress in communicating meaning through texts (i.e., productive language) that attend to a specific purpose and audience. For the final time, we illustrate this with specific examples from learning activities planned for the Ecosystem Interactions and Resources unit. While reading through these examples, consider the following questions for your own units:

- What will emergent bilinguals be doing by the end of the unit to show how they can explain and apply the big idea through texts that attend to a particular purpose and audience?
- How will productive language opportunities tie into the anchor phenomenon and text?
- How will you support and scaffold emergent bilinguals' productive language throughout the unit?

PLANNING A PRODUCTIVE LANGUAGE PROGRESSION

In any unit, it is critical that emergent bilinguals not only learn to interpret texts, but also learn to produce texts that are specific to the discipline and communicate big science ideas. Ecosystem Interactions and Resources focused on the Common Core ELA writing standard *Write informational and explanatory texts* as the endpoint of the productive language progression (See Figure 11.1). To write an explanatory text for this unit, students needed to conceptually understand how ecosystems interact and how natural processes and human activity might influence those interactions. Students also needed to know the basic structure of scientific explanations, such as the commonly used structure of claim, evidence, and reasoning (CER). A written scientific explanation is a type of discourse since it involves writing for a particular purpose and audience (i.e., a tool that conveys meaning). And to be clear, scientific explanations do not just state information. The primary purpose is to explain a process, mechanism, or causal relationship to account for the unit's anchor phenomenon. You can support student understanding

Figure 11.1. Productive Language Progression

Checkpoint #1	Checkpoint #2	Checkpoint #3	Endpoint
Use familiar language to communicate patterns in data and main ideas in a scientific article.	Write sentences that describe relationships or patterns (i.e., analyzing) found in a graph.	Write sentences that interpret a pattern or relationship found in a graph using content knowledge found in relevant sources. Use nouns and noun phrases to expand ideas and provide more detail about content knowledge.	Write a cohesive text that fully explains a phenomenon supported with multiple related and relevant sources of evidence.

of the CER structure as students conduct investigations, read texts, and engage in other learning activities during a unit. For example, you might use a graphic organizer to capture a claim and evidence from an investigation supporting that claim.

Besides knowing and following the overall organization of a scientific explanation, students need to learn and use language tools that organize particular discourses (i.e., syntax). For example, students can learn to say or write sentences that help distinguish between claim (e.g., I claim that ___) and evidence (e.g., _____ is evidence that supports the claim). These sentence frames often work best for scaffolding at the level of syntax, rather than discourse or language function.

Students also need to learn to elaborate and provide more details about their claim, evidence, and the reasoning behind the evidence. Quinn et al. (2012) state how "the level of explicit detail of observation and explanation required by science and engineering is not common in everyday experience; it demands a comparable level of precision in language use" (p. 4). Throughout the unit, you can leverage all of the languages and registers that students bring to help them add more vocabulary, including key science vocabulary, to their linguistic repertoire. This expanded vocabulary allows emergent bilinguals to strengthen the clarity of the writing by adding definitions and details.

SETTING UP FOR PRODUCTIVE LANGUAGE THROUGH INTERACTIVE SCIENCE NOTEBOOKS

We suggest that you ask students to keep all of their work in some version of an interactive science notebook (ISN). Interactive science notebooks can be physical (e.g., a composition book) or digital if the classroom primarily uses digital resources (Paek & Fulton, 2021). Like many other tools described, students benefit when they have clear expectations and models for how to set up and use science notebooks. Often, interactive science notebooks include a table of contents at the start. Students can label pages sequentially by the actual page numbers or by assignment number. Students then write directly onto pages and, if needed, tape or glue printed handouts. Interactive science notebooks are useful for all students so that they can refer back to science ideas they learned over the course of the unit. For emergent bilinguals, interactive science notebooks become even more critical, since they provide a resource for language (vocabulary, syntax, and discourse) learned and used over time.

Interactive science notebooks also work great as an ongoing assessment tool. They are visual records of how student learning has progressed over the course of a unit and even a year. You can track and provide feedback on emergent bilinguals developing language over time and remind them of resources they have for continued growth.

We suggest asking students to use the right side of the interactive science notebook pages for handouts, data, notes, and so on (i.e., input). They can then use the left side to synthesize and reflect on their thinking (i.e., output). The left side is a great way to provide emergent bilinguals with creativity and choice in how they make sense of ideas in any format (e.g., visuals, concept maps) and in any language. We have used this left side as our assessment for a lesson. For instance, after learning about the carbon cycle, you may ask students to illustrate and explain a model of the carbon cycle. While you will be looking for particular conceptual understandings, like arrows that indicate the flow of carbon, you are also accessing a window into student thinking. How is this student making sense of the carbon cycle? What linguistic tools does the student use (e.g., writing in narrative form, depicting a storyboard, examples from lived experiences)? The final key piece of the interactive science notebook is to make it *interactive*—meaning that students use it to facilitate dialogue with you and with other students.

DEVELOPING AND USING VOCABULARY FOR PRODUCTIVE LANGUAGE

There are several key points when it comes to supporting emergent bilinguals in developing and using vocabulary for productive language. First, do not expect students to memorize and restate unit-specific science vocabulary before providing them with opportunities and support for understanding the conceptual *ideas* being targeted. Vocabulary is a feature of language that helps students communicate abstract science ideas. Students do not just make sense of the carbon cycle by reading or reciting a definition. Instead, provide opportunities for emergent bilinguals to make connections between ideas using the linguistic repertoire already available to them—in English, in a language other than English, in nontechnical registers, and in gestures.

The second key point is to focus on supporting students to use vocabulary when producing *texts*, not just producing definitions. For example, when asking students to write an analysis from data collected, you can display and discuss a pattern-noticing checklist (e.g., What is recurring/repeating? Have you referenced time? What is the unit? What is the relationship?). Or add new vocabulary words to a word wall displayed on the whiteboard. Vocabulary should be displayed in English, but you can also display vocabulary in languages other than English. Emergent bilinguals might even be able to also share the vocabulary in languages other than English. Focus on providing emergent bilinguals with a linguistic tool (vocabulary) that can assist them in communicating complex science ideas.

The final key point is not to restrict "language support" as just supporting vocabulary. As we have described, language demands also include symbols, syntax, and discourses that students use to produce texts for a particular language function. We next discuss these other language demands.

PRODUCING WRITTEN ANALYSES AND INTERPRETATIONS FROM A SCIENCE ARTICLE

During the unit, you want to support emergent bilinguals in writing sentences that pertain to the focal language functions and science practices. In Ecosystem Interactions and Resources, we wanted students to produce sentences that analyzed and interpreted the kelp forest population dynamics graph before being asked to produce complete and unified explanations.

Before producing written interpretations, we wanted students to "write sentences that convey *relationships and patterns* found in data and/or texts." To support this second checkpoint focused on analysis, you can provide or brainstorm with students certain words and phrases that might be useful in turning ideas into complete sentences. We provided sentence frames to model a common sentence start when analyzing:

In 1950, _____.
1n 1970, _____.

One emergent bilingual responded with, "In 1950s the sea otter populaton decreased, causing ther prey (the sean urchins) to increase, eatiing a lot of kelp." In this response, the emergent bilingual is already using some cause-and-effect language, which could be expanded in revised writing.

For us, it was important that students distinguish between the function of analyzing and the function of interpreting data. For example, we discussed differences with students as we displayed a basic description of each term in English and in Spanish.

> *Teacher:* We are going to think for a few minutes about what is the *difference* [emphasized] between *analyzing* [emphasized] data and *interpreting* [emphasized] data. So let's look at analyze for a minute. To *analyze* [pointing to projected definition] is to examine something carefully. To look at, examine, think about something carefully. That is what you and your partner just did on the whiteboard. Were you looking and thinking carefully at this graph?
>
> *Students:* Yeh.
>
> *Teacher:* Absolutely. On those whiteboards, you wrote some really interesting ideas about what you were examining closely. Now, it's all about interpreting. *Interpreting* is about explaining the meaning of something. On your whiteboards, a lot of you said, "In 1950, the sea otter population went down [gesturing down with hand]—or decreased. In 1970, we noticed it started to increase [gestures up with hand]. That was your analysis. Now, you are going to think about your interpretation. *Why* is it going down? How do we *explain* this pattern?

In this exchange, we are building on students' own ideas and helping them see how analyses function to note patterns and trends and interpretations function to explain those patterns and trends using science principles, mechanisms, theories, and so on.

We then asked students to work with their partner and add to their analysis so that they are *interpreting* the meaning of patterns and trends found. You can provide different sentence frames that support producing interpretations (see Figure 11.2 for an example). You can also remind or discuss with students what part of a science article or investigation would be particularly useful when writing their interpretation.

For example, we asked student pairs to reread a particular passage from the "Ecosystem Superheroes" article. As one partner read a paragraph, the other partner listened and underlined any instance of a natural process or human activity. Students began to apply their understanding of these two vocabulary terms to better interpret the article, which reinforced the vocabulary. Once students identified natural processes and human activities from the text, student pairs then collaborated to write a response to the higher Level 2 inquiry question on a whiteboard:

Based on this graph and new evidence from this text, what **natural processes** and **human activity** do you think may be affecting this **ecosystem** and **its resources**? (Add potential causes to different points on the graph where the population size increases or decreases using evidence from the text.)

Figure 11.2. Sentence Frame for a Written Interpretation

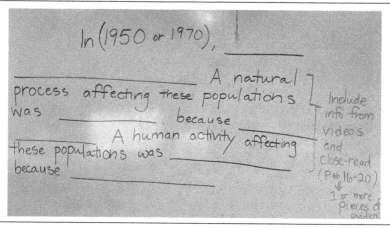

We asked students to write additions or revisions below their original ideas that were elicited earlier in the unit. Students might first add potential causes to different points on the graph where the population size increases or decreases using evidence from the text. They can add evidence from the text and other sources that support the patterns first described.

When calling upon emergent bilinguals to produce language, you want to ensure that emergent bilinguals can use their full linguistic repertoire. This means that they can use any language that is available to them as well as different modes (e.g., written and oral), modalities (e.g., print, multimedia), or models (e.g., diagrams) For instance, students might supplement written sentences with a diagram, storyboard, or flow chart, or start with the visual depiction before writing sentences.

PRODUCING GRAPHS, DATA ANALYSES, AND DATA INTERPRETATIONS

We also wanted emergent bilinguals to be able to produce graphs, data analyses, and data interpretations from an investigation. We wanted to ensure that this collaborative inquiry was meaningful and relevant, or contextualized. Based upon students' own interest in natural places around Sonoma County, we decided to take students to Shell Beach—about 45 minutes away from the school—which represents the kelp forest ecosystem. However, since we planned the unit during the COVID pandemic, we decided to take students on a virtual field trip to Shell Beach instead. We began the lesson with a land acknowledgment recognizing that this ecosystem is in the original homeland of the Southern Pomo and Coast Miwok people. You can find tips on how to create an indigenous land acknowledgment through resources such as the Native Governance Center (see https://nativegov.org /news/a-guide-to-indigenous-land-acknowledgment).

We then introduced students to the geography and history of Shell Beach, using a video of a tide pool at the beach that included focal organisms/populations living in this ecosystem (ochre sea stars, California mussels, goose barnacles). Students made observations and estimated the number of ochre sea stars, California mussels, and goose barnacles found in a 1m² area from two locations at Shell Beach. Location 1 was a rock that experiences considerable wave action and is sheltered by a rock ledge above. Location 2 was the same height from the sand, but it was drier and more exposed (see Figure 11.3). Students shared estimates and arrived at consensus estimates for each population count. Together, we and students developed our understanding of the term *population density*, reviewed the formula, and calculated together for Location 1. Independently, students calculated population density for Location 2.

Beyond gaining skills on data collection and density calculations, the purpose of this investigation is for students to deepen their understanding of ecosystem resources and interactions while also learning to communicate analyses and interpretations from an investigation. We also wanted to help

Figure 11.3. Data Collection Instructions

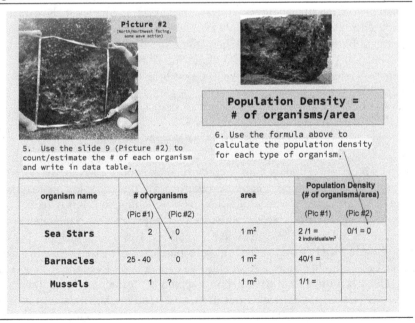

organism name	# of organisms		area	Population Density (# of organisms/area)	
	(Pic #1)	(Pic #2)		(Pic #1)	(Pic #2)
Sea Stars	2	0	1 m²	2 /1 = 2 individuals/m²	0/1 = 0
Barnacles	25 - 40	0	1 m²	40/1 =	
Mussels	1	?	1 m²	1/1 =	

students learn to make decisions about how to record, graph, analyze, and/or interpret. In this case, students decided on what type of graph (bar vs. line) would best visually represent the population density data. You could ask students to generate graphs in their interactive science notebook, on a shared poster, on the whiteboard, or on a Google Slide.

We then asked students to analyze data through three guiding questions:

- How are population densities different/the same in the two rock locations? (Is there a pattern in the distribution of organisms? Explain.)
- What factors do you think are affecting population density in the two areas? Explain.
- What resources are important to these populations? What happens if we add or remove these resources?

You can encourage students to flip back to prior work in their interactive science notebook (ISN) that will be useful for the new writing. For instance, in prior lessons the class had collectively defined and identified examples of ecosystem resources. Now, when asked in the third question what resources are important and what might happen if added or removed, students flipped to these ISN pages for support. In addition, the interactice science notebook is where students had recorded key vocabulary, descriptive words, and relevant phrases (e.g., affect, if added . . . affect things) that they could use when producing their analysis and interpretation. Figure 11.4 shows an example from part of a Google Slide produced by a small group with multiple emergent bilinguals. We helped them add more details and encouraged them to record any questions they had about the resources. We also encouraged them to write ideas in Spanish for the first time before practicing writing in English.

SUPPORTING "HOW LANGUAGE WORKS"

So much of our language and literacy focus has been on supporting emergent bilinguals in interacting in meaningful ways (i.e., *how* language is used) and using language purposefully (i.e., *why* use language). But there are also opportunities in the science classroom to support how language works (i.e., *what* linguistic resources are available). Resources include, but are not limited to, (1) using verbs and verb phrases, (2) using nouns and noun phrases, (3) modifying to add details, (4) connecting ideas, and (5) condensing ideas. It is important to note how "spelling," "grammar," and "punctuation" are *not* explicitly called out in this list. You might model and provide feedback on emergent bilinguals' spelling, grammar, and punctuation, but it

Figure 11.4. Student Responses to Resources for Populations

LH 21 #3 What <u>resources</u> are important to these populations? What happens if we add or remove these resources?

- The rock – provides **SHELTER** (home, protected place to be)
  - How does smooth or bumpy sections affect?
  - How does different colored areas affect?
  - What areas are better for hiding and moving if you are a sea star or mussel?
  - Which side/area gets more sunlight and water/waves?
- Rock – provides **minerals** they need
- **FOOD**
  - Clams, Sea stars and Seaweed
  - Mussels – food for the sea stars
- Water – necesita más agua para vivir
- Oxygen

---

should be in the service of helping emergent bilinguals know how to structure clear and cohesive texts, in other words, the linguistic resources with which to communicate meaning for particular purposes.

In *Ecosystem Interactions and Resources* we capitalized on opportunities for emergent bilinguals to learn to use nouns and noun phrases to expand ideas and provide more detail. At the beginning of the unit, students were expected to *use familiar language to notice and communicate patterns in data.* Closer to the end of the unit, they were expected to *write multiple sentences that convey main ideas (or claims) and multiple related and relevant sources of evidence.* Native English speakers might still be working on crafting sentences that explain or justify an idea.

You can help students draw on prior knowledge to begin the sentence with a claim (i.e., a particular interaction pattern) followed by some reason or example (i.e., "because" or "for example"). However, they will have the advantage of some repertoire of nouns (i.e., person, place, or thing) that provide the subject or object of the idea (e.g., scientists can analyze *predator–prey interaction patterns*). When students are expected to write lengthier, more complex scientific explanations, they will need to go beyond just naming the interaction and instead describe it in richer detail. This is where you can help students brainstorm what nouns could be useful to add more detail. The class might come up with the *predator–prey interaction patterns* and then the teacher could use a spider diagram to challenge students to come up with as many details as possible about the *predator–prey interaction patterns*, as shown in Figure 11.5.

Figure 11.5. Spider Diagram for Nouns and Noun Phrases

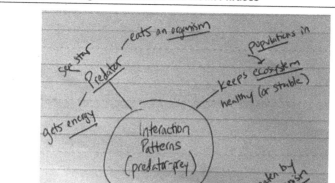

We suggest targeting support for how language works toward the end of a unit as students are working toward even clearer communication of science ideas to analyze, explain, or justify ideas. As students' learning deepens, so can their ability to use evidence to reason about what is happening in ways that use language conventions (like noun phrases) to expand ideas with a larger repertoire of vocabulary. You can facilitate "gallery walks" (i.e., displaying written work on the whiteboard or posters taped to the wall as students "walk" and read the work) or ask students to exchange sentences with a partner to get feedback and ideas on ways to add detail. The key is a meaningful goal and activity that are integrated with learning and doing science, not separate from it.

DRAFTING, REVISING, AND SHARING TEXTS

The culminating productive language expectation for Ecosystem Interactions and Resources is for students to write a cohesive text that predicts patterns of interactions among organisms across multiple ecosystems. Everything emergent bilinguals have been noticing, wondering about, analyzing, and interpreting during the unit prepares them to communicate meaning through a text. Along the way, emergent bilinguals have developed a deep conceptual understanding of how ecosystems are organized and what natural processes and human activities influence kelp forest ecosystems. Now they can apply this conceptual and linguistic understanding to their own chosen organism.

This is also an opportunity for emergent bilinguals to make meaningful connections to their own out-of-school lived experiences. As an example, one student shared how important hawks and hummingbirds were to his family and culture:

> In our native american tribe (coast miwok), males after they die are red tail hawks and females are hummingbirds. In the valley in between sonoma mountain and taylor mountain there is a lot of hawks everywhere.

One reason this is important is so that students enter into writing through personal experience or background information research, but eventually fully explain the science.

The final product itself, the blog, is important to demonstrate what they can do. But the process of drafting, revising, and sharing the blog is also important. One of the Common Core ELA Writing Standards is for *students, with some guidance and support from peers and adults, to develop and strengthen writing as needed by planning, revising, editing, rewriting, or trying a new approach, focusing on how well purpose and audience have been addressed.* An important aspect is also considering the audience when choosing what language will best communicate these ideas.

We provided ongoing support for students to draft, revise, and share their work. For instance, we planned guiding questions for each blog post to help students think and communicate deeper, more complex ideas and applications. These guiding questions work the same way as leveled inquiry questions for interpreting the anchor text. Figure 11.6 shows the prompt and a sample student response. In the response, the teacher can notice how in the first blog post this student directly answers each prompt, but now can work on elaboration and making transitions between sentences. However, by the second blog post there is both evidence of conceptual understanding and explanation of idea ("The reason is that . . ."). The next step for this student could be to cite particular sources and/or provide more direct evidence.

You and your students can use the evaluative criteria formatively—to provide feedback as students draft posts. Before using criteria, provide a space for students to read and make sense of the criteria—both individually and collaboratively. For example, you might share the criteria, and ask students to annotate just like they would a text (e.g., key ideas, confusing words). You can also provide example blog posts that differ in how well they meet the criteria. You can even ask students to discuss in groups how they would evaluate the examples and why. These supports let any confusions about the criteria surface, such as exactly what does it mean to "extend the student's thinking." You can then ask students to use the criteria to peer-review each other's work and/or self-assess their own work.

Figure 11.6. Sample Blog Posts

Ecologist Research Blog Post #1 *Rough Draft*

ORGANISM FUN FACTS! (Part 1)

Directions:

Write a paragraph in the box below (4 or more complete sentences) that includes the following information:

- What organism are you researching and sharing about (including its common and scientific name)?
- Where does your organism live (where in the world and what ecosystem/habitat type)?
- Why did you choose to learn more about this organism? What are 2-3 fun, interesting or exciting things to you about this organism?

**Use ISN 20, ISN 21, and Intro to Ecologist Research to help you as needed!*

> The organism that I am researching is the Blue-Eyed Darner. The Scientific name used for this organism is **Rhionaeschna multicolor.** The Blue-Eyed Darner can be found in lakes, ponds, marshes, and swamps. They are found in the west of North America, east of Wisconsin, and South of Texas. Its ecosystem type is the Coastal Prairie. I chose The Blue-Eyed Darner because I hadn't heard of them before. I learned that they could fly up to 50 mph. I also found that they have 360-degree vision.

ORGANISM FUN FACTS! (Part 2)

Directions:

Write a paragraph in the box below (4 or more complete sentences) that includes the following information...

- Is your organism **essential to the ecosystem/habitat** it lives in? *(What might happen to that ecosystem if your organism wasn't there?)* Why?
- What RESOURCES does it use and need in its habitat?
- What LIMITING FACTORS impact the population growth of your organism? Explain!

**Use ISN 20, ISN 21, and Intro to Ecologist Research to help you as needed!*

> The Blue-Eyed Darner is important to the aquatic ecosystem that it lives in. If The Blue-Eyed Darner wasn't there, I think that there would be a lot more mosquito larvae and that might mess up lots of land or water, aka, overpopulation. The reason is that the Blue-Eyed Darner is the insect that eats mosquito larvae. The Blue-Eyed Darner also eats other aquatic larvae. The Blue-Eyed Darner's most needed resource is water. They are born on the water and their kids are born on the water as well, so they need rivers and ponds and other bodies of water. It is also sad because Blue-Eyed Darners are dying because there is becoming less freshwater worldwide, which is not good for their babies as well since they are born in water, not to mention that their habitat,(water) is in very poor condition in the U.S, just to add on to their problems. I think that is the only resource they need. Though other important factors are their food, their food is aquatic larvae, and if the Blue-Eyed Darner was not near a body of water, or it can't find one how is it going to eat? Also, Blue-Eyed Darners are in danger because there is too much nitrogen and phosphorus due to pollution, and it is just too much for them.

CONCLUDING PRODUCTIVE LANGUAGE REMINDERS

- **Position Students as Sensemakers:** Design assessments that allow emergent bilinguals to apply the core ideas, crosscutting concepts, and science practices learned during the unit.
- **Facilitate Productive Student Discourse:** Provide opportunities for emergent bilinguals to share and discuss texts that are produced.

- **Scaffold Language and Literacy:** Ensure that students use information to produce texts that have a specific purpose and audience. Provide sentence frames, graphic organizers, or other supports, but work toward helping students need these supports less and less over time.
- **Contextualize Science Activity:** Connect the culminating assessment to the anchor phenomenon and text. Give students choice of what or how they produce and share ideas. Base the anchor text on known student interest or out-of-school, lived experiences. Keep coming back to the anchor phenomenon.

Conclusion

Strengthening Science Instructional Planning for Emergent Bilinguals Through Collaboration

We wrote this book to provide you (science teachers and those working with science teachers) with tools and a theoretically grounded approach to plan science instruction for emergent bilinguals that weaves in rich and relevant language support. Without any guidance, that is a daunting task for experienced science teachers, science teacher educators, and certainly those beginning their journey into teaching. We are not the first to compile strategies to add to your repertoire of how to teach emergent bilinguals. But what we felt was important is a renewed, comprehensive focus on intentional planning.

Teaching emergent bilinguals is not about eliminating the language that makes science learning hard or making minor adjustments to general instruction, such as adding sentence frames or pictures. In the approach we articulated, you can zoom out and consider the entire, integrated conceptual and linguistic landscape. What's the big science idea to explore and explain? To do this, what would students be doing with language? When would they be interpreting? Collaborating? Producing? How will they develop the discipline-specific vocabulary, syntax, and discourse so that they can use language in these different modes for particular purposes? What do I even know about my emergent bilinguals that I can leverage, and how can I elevate all of the languages and lived experiences they bring into the classroom? We hope the foundational research, planning tools, model unit, and specific examples make the planning process more accessible and provide some inspiration that yes, you can do this!

Science teachers spend their entire elementary, secondary, and postsecondary education (potentially including graduate work) developing a deep understanding and appreciation for science. But some science teachers only get a year (or maybe a few if they begin in an undergraduate program in education) of intensive teacher preparation. And then how much of that time is devoted to really understanding what it means to be an emergent bilingual and learn science in today's educational context, and what does it mean to develop one's language and literacy? It will be extremely valuable to have more science teachers who share critical identities with the students they serve—meaning more science teachers who speak a language other than English (in the United States) or better yet, are emergent bilinguals themselves. But that is not a requirement. Monolingual teachers (and that applies to both authors of this book) can still find out about the lived experiences and the linguistic and cultural resources of their emergent bilinguals. Using this awareness of language development and the language being asked of students through science and engineering practices, even monolingual teachers can carefully plan learning activities and whole units that support language and literacy (specifically) and are responsive to emergent bilinguals more generally. We have used SSTELLA's four instrumental dimensions as a mantra to keep reminding ourselves of what to consider when framing an entire unit and planning specific learning and assessment activities. They are worth mentioning again:

- Position all students as sensemakers.
- Facilitate productive disciplinary discourse.
- Scaffold language and disciplinary literacy.
- Contextualize instruction.

But we do not expect you to do all of this alone. Therefore, we conclude by encouraging you to collaborate with others so that expertise and experiences can be shared, discussed, and used to support teaching in science and across all the subject areas.

COLLABORATION BETWEEN SCIENCE TEACHERS AND UNIVERSITY RESEARCHERS/TEACHER EDUCATORS

This book is the direct result of a reciprocal and synergistic relation between a science teacher and a university science teacher educator—a relationship similar to that between science and language/literacy as articulated throughout the book. As a university science teacher educator, I (Edward Lyon) secured and worked on grants that allowed mentor science teachers to engage in professional learning around teaching emergent bilinguals alongside preservice teacher candidates. In the process, mentors such as Kelly Mackura could share practical expertise to ensure that the theoretically rich teaching practices promoted through grants and initiatives worked in particular student and school contexts. This collaboration allowed Mackura to develop into a teacher leader and guide her colleagues in elevating their own teaching and mentorship. What was essential in this collaboration was a common goal—supportive and responsive science teaching for emergent bilinguals. We brought different experiences, areas of expertise, and perspectives that enriched each other's work. Our work together could be further enriched by including biliteracy experts who can help both science teacher and science education researchers and teacher educators reflect on teaching and put *bilit-eracy*, not just *English* language development, front and center.

COLLABORATION BETWEEN SCIENCE TEACHERS AND ENGLISH LANGUAGE ARTS AND ENGLISH LANGUAGE DEVELOPMENT TEACHERS

We also encourage collaboration between you and English language arts (ELA) and English language development (ELD) teachers at the same school or district. For one, the ELA and ELD instructors can provide further insight into the prior learning and targeted goals for emergent bilinguals that might be shared in class. You can use common supports, such as the annotation guide. You can also use common texts, especially since ELA teachers are also supporting students in reading and writing informational, explanatory texts. What a powerful experience if emergent bilinguals analyzed the "Ecosystem Superheroes" article in their ELA class to focus more on linguistic choices and style, whereas you can delve more into using the ideas and evidence to support explaining ecosystem interactions. Even better, the article can be read and analyzed in Spanish (or other home language) to foster students' biliteracy.

COLLABORATION BETWEEN MENTOR SCIENCE TEACHERS AND SCIENCE TEACHER CANDIDATES

The planning tools and model unit are meant to be used by science teachers of all levels—from preservice to experienced. There is an added benefit to preservice teachers when they coplan alongside their school-site mentor (i.e., cooperating or master) teacher. Just as we did throughout the book, early in the year the mentor teacher can talk through and model the process of identifying NGSS PEs, ELA/ELD standards, an anchor phenomenon, detailed lesson plans, and so on. Later in the year, they may plan a new unit together, where preservice teachers can offer ideas learned in their coursework and have the flexibility to try it out in practice with feedback and debriefing with the mentor. Preservice teachers often have to plan and be assessed on their own set of lessons, which can follow the planning tools and lesson plan template used in this book. The Ecosystem Interactions and Resources unit was in fact planned with input from a preservice teacher and benefited from the perspectives they bring. Two of the units featured in Appendix B were also developed by preservice teachers who learned through many of the same tools discussed in this book. We encourage constant reflection and probing—asking preservice teachers about their planning considerations and decisions. The guiding questions from each planning tool can help here.

We also encourage looking out for evidence of the language and literacy supports as they are carried out in practice.

COLLABORATION BETWEEN SCIENCE TEACHERS AND FAMILIES/ CAREGIVERS AND THEIR STUDENTS

We can ideally bridge these collaborations between communities and form what is often called a professional learning community. And in these collaborations, we encourage the use of family/caregiver voices. While you will do your best to know about our emergent bilinguals and plan science-, language-, and literacy-rich and relevant science instruction for them, the students and their families are the ones whose education is at stake and should always be at the center of learning and of instructional planning. As described in Chapter 4, surveys and communications sent to families/caregivers can solicit relevant and rich information to inform teaching. Upon learning from families/caregivers, you can also invite them into the classroom to share experiences and expertise they may have. Alternatively, students and families/caregivers can help provide input on the unit developed or coplan it. You can invite families/caregivers in person or virtually as students present final products—such as the Ecologist Research and Blog. This gives families/ caregivers a chance to not just learn, but comment and contribute to the learning process.

Finally, emergent bilingual voices should not be forgotten and should be what we hear loud and clear—in any language. You can listen to emergent bilinguals to inform how they plan, listen as they are learning and developing language during instruction, and continue to listen and solicit feedback on how the activities supported or did not support their learning. Similar to families, emergent bilinguals can be collaborators in planning the real-world issue, text, and even assessment that frame the unit. Given that this book is ultimately in support of emergent bilinguals, it is fitting to end by hearing from an emergent bilingual who experienced the Ecosystem Interactions and Resources unit.

In the ecosystem research and blogging carried out during the model unit, one newcomer emergent bilingual student chose to research a squirrel because she connected with a squirrel like a pet when she was a young girl living in Mexico. She was encouraged to communicate her thinking and learning in Spanish, as she began developing English. In a blog post she shared:

Me gusta este organismo porque es lindo y si en algún momento quiero un ardi-lla. Necesito saber sobre lo que hacen en la naturaleza y que les gusta. Yo tuve una de chiquita y la tuve hasta que creció y la saque al corral de mi abuelita y se fue. (I like this organism because it is cute and if at some point I want a squir-rel. I need to know what they do in nature and what they like. I had one when

I was little and I kept it until it grew up and I took it out to my grandmother's corral and she left.)

Buxton and Caswell (2020) remind us that the "language of science is not about replacing students' colloquial language with scientific language, but rather about developing a broader linguistic repertoire that allows students to make nuanced linguistic choices based on their communicative purpose while also developing grade-appropriate meanings" (p. 573). The blog was a powerful translanguaging moment in class where she constructed meaning from a broader set of linguistic resources—in Spanish and while using a mix of technical vocabulary (organism) and less technical (cute). But in this telling of a memory there is also the curiosity of a scientist who wonders "what they do in nature."

Multilingual development was validated and normalized in the learning community. When she participated using her own linguistic repertoire in the classroom learning spaces, she also increased multilingual engagement among her peers. They readily responded to her and added thinking in Spanish. By intentionally planning with and around student and family assets, it became clear that those powerful moments arise where you get to explore how the relationship between science and language (and language assets of our students and families) can really support the raciolinguistic norms and assets of our students and families in important ways (Buxton & Caswell, 2020). Skillful teachers can elicit students' ideas and create these powerful moments on the spot. But we would argue that they are much more likely to arise and be leveraged through careful planning when you know the type of language demands expected of students, have a road map for scaffolding students' science learning and language/literacy development, and create the structures, norms, and other supports that will continuously foster student engagement, learning, and agency to share thinking using a range of language choices. This is what it means to plan science instruction for emergent bilinguals that weaves in rich and relevant language support.

Sample *Ecosystem Interactions and Resources* Lesson Plans

Analyzing and Interpreting the Population Graph

Lesson Focus	
NGSS PE: MS-LS2-1. Analyze and interpret data to provide evidence for the effects of resource availability on *organisms and populations of organisms in an ecosystem.*	**ELA/ELD Standards:** **CCSS ELA:** Cite specific textual evidence to support analysis of science and technical texts (interpretive). **ELD:** Interacting via written English—print and multimedia (collaborative).
Focal Science or Engineering Practice(s): Analyze and interpret data	**Focal Language Function(s):** Analyze

Integrated Learning Objective: Students will be able to analyze and interpret data in a population graph to predict patterns for how sea otter, kelp, and sea urchin populations interact and use resources in their ecosystem.

Sequence of Opportunities for Learning Science and Using Language

1. The teacher visually introduces students to organisms of interest in the population graph (*sea otter, kelp*, and *sea urchins*) and their kelp forest ecosystem/habitat through labeled images and video clips.
2. Students set up their interactive science notebook (ISN) for the unit.
3. The teacher models and elicits student noticings about line graph features and functions using the kelp forest population dynamics graph.
4. Students develop shared understanding around how to identify *patterns.*
5. Students individually think about what they *notice* and *wonder* about the population graph and write responses in their interactive science notebook.

   Guiding Questions:

   • What do you *notice* about the population graph? (What *patterns* do you see? What patterns do you notice between . . . ?) *(Round 1)*
   • What do you *wonder* about the populations in this graph?
6. Students collaborate and share thinking around population graph notices and wonderings via science talk in small groups.

(continued)

7. Teacher records and discusses group notices and wonderings on a slide, poster, or whiteboard, shared with the larger class.

8. Steps 5–7 are repeated with students thinking about the larger question framing the unit: *Based on this graph, what natural processes and human activity do you think may be affecting this ecosystem and its resources?*

Discipline-Specific Language Supports

Interpretive	Collaborative	Productive
(making sense of oral and/or written language)	(communicating with others to make meaning)	(communicating meaning through oral and/or written language)
• *Video: The Sea Otters Enchanted Forest* to better contextualize and visually introduce students to the anchoring phenomenon.	• *Science talk* norms/ expectations/group roles are visible, modeled, and practiced.	• *Interactive science notebook* (ISN) to allow students to refer back to prior resources, language used, and learning, as well as track learning progress.
• *Patterns noticing checklist* to support knowing how to look at patterns in a graph.		• *Sentence frames* available to all students to support construction of sentences that analyze the graph.
• *Guiding questions* to support how students begin to read and think about a graph along with opportunity to identify line graph structures and purpose.		• *Group poster/white board/poster* to allow students to produce and share authentic analysis ideas with multiple modes/ modalities.
• *Time to think and write* so that students can process and individually interpret what they have been reading and hearing.		

Citing Evidence from the "Ecosystem Superheroes" Article

Central Focus

NGSS PE:	ELA/ELD Standards:
MS-LS2-2. Construct an explanation that *predicts patterns of interactions among organisms across multiple ecosystems.*	**CCSS ELA:** Cite specific textual evidence to support analysis of science and technical texts (interpretive language).
	ELD: Reading closely and explaining interpretations and ideas from readings (interpretive language).
	ELD: Interacting via written English—print and multimedia (collaborative language).

Focal Science or Engineering Practice(s):	Focal Language Function(s):
Evaluate information	Analyze and interpret

Integrated Learning Objective: Students will be able to cite specific evidence from a text (close reading) to explain patterns for how sea otter, kelp, and sea urchin populations interact and use resources in their ecosystem.

Sequence of Opportunities for Science Learning and Language Use

1. Students describe roles and relationships in their own lives.
2. Students list and share what they notice about and learned from the *Exploring Ecosystems: Coastal Food Webs* video.
3. Teacher models how to read closely the article "Ecosystem Superheroes: Sea Otters Help Keep Coastal Waters in Check" (Level 1 inquiry question #1).
4. Students annotate and share annotations from the article with partner.
5. Students individually complete and then share responses to Level 1 inquiry questions #2 and #3.
6. Students (re)watch *Exploring Ecosystems: Coastal Food Webs* and describe roles and relationships found in the video.
7. Students (re)analyze the Kelp Forest Population Dynamics Graph.
8. Partners close-read article looking for "natural process" and "human activity."
9. Partners complete Level 2 inquiry questions.

Language and Literacy Supports

Interpretive (making sense of oral and/or written language)	Collaborative (communicating with others to make meaning)	Productive (communicating meaning through oral and/or written language)
• *Video graphic organizer:* Helps students organize ideas while watching the coast food webs video and make connections to their own lived experiences.	• *Purposeful grouping of partners:* Based on reading abilities in English and in primary language.	• *Sentence frames:* Scaffold students' use of phrases related to analyzing and interpreting.

(continued)

- *Close reading annotation protocol*: Establishes norms for interpreting a text.
- *Leveled questions:* Scaffold students' interpretation of a text.
- *Raise your voice!:* All important terms are pronounced by the instructor and repeated chorally by the entire class, in English and in the primary language of emergent bilinguals.

- *Collaboration board:* Provides an opportunity for students to exchange ideas via print rather than verbally.

- *Group poster/white board/poster:* Response protocol visually available for all students.

Analyzing Tide Pool Population Data

Central Focus	
NGSS PE:	**ELA/ELD Standards:**
MS-LS2-1. Analyze and interpret data to provide evidence for the effects of resource availability on organisms and populations of organisms in an ecosystem.	**CCSS ELA:** Cite specific textual evidence to support analysis of science and technical texts. Write informational and explanatory texts. **ELD:** Use nouns and noun phrases to expand ideas and provide more detail.
Focal Science Practice: Analyze and interpret data	**Focal Language Function(s):** Analyze and interpret

Integrated Learning Objective: Students will be able to accurately record, graph, analyze, and interpret population data in order to predict patterns of how organisms compete for and share resources in a California coast tide pool ecosystem.

Sequence of Opportunities for Science Learning and Language Use

1. Students participate in a land acknowledgment and are introduced (virtually or in person) to the location of interest (Shell Beach, Sonoma Coast State Park) and to the focal organisms/populations living in this ecosystem (ochre sea stars, California mussels, goose barnacles) through labeled images and video clips.
2. Students set up their interactive science notebook—including a space for observations/notices and a population count and density data table.
3. Students view **two locations on a rock. Location 1 is on one side of the rock, and Location 2 is on the other side of the same rock at the same height from the sand but drier and more exposed.** Students make observations and count/estimate the number of *ochre sea stars, California mussels, and goose barnacles* in a 1m² area at each location.
4. Students share their observations and estimates in pairs or small groups. The class agrees on consensus estimates for each population count.
5. Together, the instructor and students define *population density*, review the formula, and calculate together for Location 1. Independently, students calculate population density for Location 2.
6. Students create population density bar graphs for Locations 1 and 2.
7. Students analyze and interpret the population density data and answer the analysis questions in their interactive science notebook.
8. Students share collaborative thinking with other pairs/small groups as well as in a large-class science talk.

(continued)

Language and Literacy Supports

Interpretive	Collaborative	Productive
(making sense of oral and/or written language)	(communicating with others to make meaning)	(communicating meaning through oral and/or written language)

Interpretive (making sense of oral and/or written language)

- *Contextualizing language use* through video, pictures, and historical context.
- *Bar graph reference* (anatomy and purpose) available to students to support understanding of x axis, y axis, bar creation, use of color, and how to compare bars.
- *Raise your voice!—* All important organism and location names are pronounced by the instructor and repeated chorally by the entire class.

Collaborative (communicating with others to make meaning)

- *Group roles* assigned to each collaborative group as they collect and analyze data

Productive (communicating meaning through oral and/or written language)

- *Interactive science notebook* assignment(s): available for students (in handout, digital format and/or interactive science notebook entry) (Discourse/Syntax).
- *Group poster/ whiteboard/poster:* Response protocols visually available for all students—observations/ data collection, bar graphs, discussion questions (Discourse/ Syntax).
- *Noun and noun phrases spider diagram:* to add details to writing.

Additional Unit Frames and Student Resources

UNIT I—HIGH SCHOOL BIOLOGY: TESTING FOR CANCER THROUGH DNA MICROARRAY TECHNOLOGY

Contribution by Suzanne Garcia

Unit Frame

Essential Student Assets

- **Personal interests:** Students are active community members; many work or volunteer with local community-based organizations. Athletic culture is very strong as well, with basketball being particularly popular among students.
- **Linguistic repertoire:** Approximately 2% of our student body are designated or redesignated English learners. For our emerging EL students we offer two sections of Academic Workshop Support and an English language development class, levels 1–4. As a result of our small numbers of EL students, math and science teachers have minimal systems of support within their classrooms for these learners. Among our emergent bilinguals the most common languages include Spanish, Gujarati, Urdu, and Hindi.
- **Home/family/community assets:** Our students' assets are unique in that we have a significant range in socioeconomic status, English language proficiency, and community composition. The majority of our students come from local affluent suburban communities, but we also serve students from rural/agricultural communities and public housing.
- **Out-of-school science experiences:** Similar to family assets, our students show a tremendous range in out-of-school science experiences. Select students participate in high-level extracurricular STEM activities like robotics, ecological restoration, science camps, and wildlife rehabilitation' while others have little to no access to any science enrichment.

Big Science Idea: The structure of DNA allows it to code for information to make proteins.

(continued)

Essential NGSS Standards:	Targeted CCSS ELA and ELD Standards
Construct an explanation based on evidence for how the structure of DNA determines the structure of proteins that carry out the essential functions of life through systems of specialized cells.	**Interpretive:** Follow precisely a complex multistep procedure when carrying out experiments, taking measurements, or performing technical tasks, attending to special cases or exceptions defined in the text. **Collaborative:** Exchanging information/ideas via oral communication and conversations. **Productive:** • Write informative/explanatory texts, including the narration of historical events, scientific procedures/experiments, or technical processes. • Select and apply varied and precise vocabulary and other language resources.
Anchor Phenomenon: Students take on the role of an oncologist (a doctor who studies and treats cancer) who needs to determine the gene profiles in newly diagnosed skin cancer patients.	**Anchor Text:** "How DNA microarrays work" [see student resource below]

Culminating Assessment Activity: Students write an evidence-based explanation to account for what gene or genes they think are responsible for the cancer of Patient 1 or Patient 2.

Students are given two options for how to share their work:

• Option 1: Students will draft an outline of their CER on a large whiteboard (1 claim, 1 piece of evidence, and 1 sentence of reasoning). Then students will engage in a gallery walk to look for similarities and differences among drafts. These can include similarities and differences in data cited or differences in sentence structure. For example, "I notice claims are written as statements."
• Option 2: The teacher can gather draft claims, evidence, and reasoning, either physically or in a Google form or Canvas survey. The teacher can then post the results and students can work collectively to rank the best claim, evidence, and reasoning. A class discussion about why certain choices appear strongest can support sensemaking while students formulate their own CERs.

Targeted Language Demands

• **Language functions:** Explain.
• **Vocabulary:** gene, transcription, translation, gene expression, mutation.
• **Syntax:** Phrases that transition between and coordinate claim-evidence-reasoning (e.g., my evidence is, for example, in conclusion).
• **Discourse:** Claim–evidence–reasoning structure.

Student Resource: How DNA Microarrays Work

Gene Expression

Almost every cell in our body has a copy of the exact same DNA, with all the same genes. What makes each cell different are which genes in that cell's DNA are being **expressed** (turned on), or **silenced** (turned off). Which genes get expressed in your muscle cells are different from the genes expressed in your skin cells. Why? Because skin and muscles have different jobs, and require different proteins.

Using Microarrays to Learn About Cancer

Two ways that cells can become cancerous are when too much of a protein is made, or when a protein is made incorrectly. Genes often work as a team to carry out the functions of the cell. One gene may even control another, determining when it gets expressed and how much protein it makes. Think about a gas pedal and a brake pedal in a car. When you go too fast, you can slow down with the brakes. Sometimes genes will change or **mutate** because they are copied incorrectly. A change in the DNA instructions may mean that the protein is made differently, and works differently. If either of these changes take place with proteins that regulate the cell cycle, they can trigger the cell to divide uncontrollably, causing cancer.

This is where DNA microarrays can help. Doctors can perform a DNA microarray analysis to determine which genes are being expressed or silenced in healthy cells versus cancerous cells. By identifying which genes in the cancer cells are working abnormally, doctors can better diagnose and treat cancer.

We know from **transcription** that when a gene is expressed in a cell, it generates messenger RNA (mRNA). In a microarray a DNA copy of the mRNA is made and analyzed.

Steps in Setting Up a DNA Microarray

A real microarray slide is about the size of a postage stamp, with thousands of tiny wells to test thousands of genes. In our lab we will use a spot tray with six wells to test six genes.

1. **Collect tissue.** The first step is to collect healthy and cancerous skin tissue samples from the patient. This way, doctors can compare what genes are being expressed.
2. **Isolate mRNA.** Once the tissue samples have been collected, the mRNA is extracted from (taken out of) both the healthy and cancerous tissue samples.

3. **Make a labeled DNA copy.** A DNA copy is made from both the healthy and cancerous mRNA (think of this step like backwards transcription). This DNA copy is called complementary DNA (cDNA). Each cDNA is labeled using fluorescent nucleotides. The cDNA from the healthy cells glows *green*, and the cDNA from the cancerous cells glows *red*. We make a cDNA copy because DNA is more stable than RNA and is less likely to break apart.

4. **Apply cDNA.** The green and red labeled cDNA are mixed and added to the microarray.
5. **The cDNA binds.** The cDNA will bind to any well on the microarray that has a matching sequence of synthetic DNA. Extra cDNA that did not bind is washed off the microarray. In our lab, each well contains a synthetic strand of the following genes: CDK4, CDKN2A, MAGEA1, MC1R, ATP5J, and HBB.

What Do the Colors Mean on a DNA Microarray?

Each well in a microarray **corresponds with a specific gene sequence.**

- **Red:** the gene is expressed in the cancer cell and silenced in the healthy cell.
- **Green:** the gene is silenced in the cancer cell and expressed in the healthy cell.
- **Yellow:** the gene is expressed in both the cancer and healthy cell.
- **Black:** the gene is silenced in both the cancer and the healthy cell.

Our lab is a <u>simulation</u> of a microarray, our colors will be different.

Student Resource: Using DNA Microarray Analysis to Investigate Skin Cancer

> **Purpose:** To interpret the genetic profile of each patient using DNA microarray analysis and to determine which genes may contribute to their cancer.
>
> **Background:** You are an oncologist (a doctor who studies and treats cancer). You will be performing a microarray analysis to determine the gene profiles in newly diagnosed skin cancer patients. Based upon your analysis, you will determine which genes may be causing the patient's cancer.

You will be given two trays with 6 wells. Each well contains a **known sequence of synthetic DNA** for each of the following genes CDK4, CDKN2A, MAGEA1 MC1R, ATP5J, and HBB. Synthetic DNA is made by assembling nucleotides in the lab.

mRNA has already been extracted from healthy and cancerous tissue cells from your patients, and a **cDNA** copy of the mRNA has been made. You will add each patient's **cDNA solution** to *their* microarray.

If the cDNA hybridizes (combines) with the synthetic DNA, the well will turn orange. If the cDNA does not hybridize, the well will turn green.

Procedure

1. A microarray for each patient has been prepared for you ahead of time (thanks, nurse!). Each microarray well contains equal amounts of synthetic DNA copies of each of the 6 genes listed in the microarray table.

Patient 1   Patient 2

CDK4 (Gene 1)	CDKN2A (Gene 2)	MAGEA1 (Gene 3)	CDK4 (Gene 1)	CDKN2A (Gene 2)	MAGEA1 (Gene 3)
MC1R (Gene 4)	ATP5J (Gene 5)	HBB (Gene 6)	MC1R (Gene 4)	ATP5J (Gene 5)	HBB (Gene 6)

2. Put **two drops** of **Patient 1's cDNA** in each well for **Patient 1**. Put **two drops** of **Patient 2's cDNA** in each well for **Patient 2**.

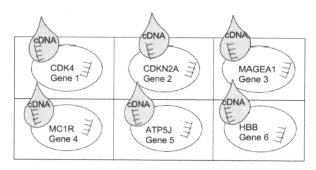

3. Use the microarray key in the results section to record which genes are being expressed or silenced. Record your results by circling the appropriate symbol in your microarray table.

Results

Circle the symbol that best represents the level of gene expression. Use the *Microarray Key* to pick a symbol that represents your results.

Microarray Key

Microarray Simulation Color	Gene Expression Symbol (+, −)	Gene Expression In cancer cell
Orange	+	Expressed
Green	−	Silenced

Microarray Results
Patient 1

Gene 1 CDK4	Gene 2 CDKN2A	Gene 3 MAGEA1
+ −	+ −	+ −

Gene 4 MC1R	Gene 5 ATP5J	Gene 6 HBB
+ −	+ −	+ −

Patient 2

Gene 1	Gene 2	Gene 3
CDK4	CDKN2A	MAGEA1
+−	+−	+−
Gene 4	Gene 5	Gene 6
MC1R	ATP5J	HBB
+−	+−	+−

Student Resource: Gene Profiles

Below is a list of genes and information about the proteins they code for. The expression or silencing of these genes can help doctors learn more about what kind of cancer a patient may have and how to best treat their disease.

Gene Name	Information and Function
CDK4 (Gene 1)	A member of the protein Kinase family (a group of similar types of proteins). Regulates the G1 and S phase of the cell cycle. When expressed, this gene is associated with tumor growth and melanoma (skin cancer).
CDKN2A (Gene 2)	Inhibits (stops or slows down) CDK4, and the production of cyclin-dependent kinase. It is known to be an important tumor suppressor protein (kills cancer cells when they first form).
MAGEA1 (Gene 3)	A transmembrane protein that regulates transcription. When expressed, it has been shown to produce a protein that creates conditions in the cell that are favorable for tumor growth. This gene is located on the X chromosome, and is associated with inherited diseases.
MC1R (Gene 4)	A receptor protein that builds skin pigment (color). When functioning properly, this gene produces a tumor suppressor protein. Variations in this gene are associated with light skin and hair color and increased sensitivity to UV radiation.
ATP5J (Gene 5)	Builds an enzyme that promotes ATP synthesis in the mitochondrial membrane. ATP is a molecule that provides the energy needed to power all cell functions.
HBB (Gene 6)	Builds part of the hemoglobin protein in red blood cells. Hemoglobin carries oxygen. Mutations in this gene are connected to sickle cell anemia.

UNIT II—HIGH SCHOOL BIOLOGY: GENETIC VARIATION THROUGH THE PHENOMENON OF WHIPTAIL LIZARDS

Contribution by Annette Bustamante and Kelly M. Mackura

Unit Frame

Essential Student Assets

- **Personal interests:** Many students are interested in domesticated animals and their personal pets—especially dogs. They are interested in breeding, differing breeds, and varying traits of their pets. Some students also work or have worked with steer breeding in their home countries.
- **Linguistic repertoire:** Developmentally, students really enjoy debating and arguing ideas. Yet they are still learning how to explore a complex idea deeply from many different perspectives. They are also learning how to listen to their peers use the language and ideas of others to build and grow their own thinking and ideas.
- **Home/family/community assets:** There are varied roles males have played in the lives of my students. Students have varying perspectives of male gender roles. Some of my students have fathers living with them or always present and others do not. This allows students to bring different perspectives regarding what it means to "need" a male (in humans and other species)
- **Out-of-school science experiences:** Students experience the benefits of genetic variation in ever-changing environments. (For example: immunity to ever-evolving pathogens, cancer, allergies, mating selection, sensory abilities, environmental protections, camouflage/escaping predators, etc.) They also have experience in the genetic variation and breeding with domesticated animals and pets in their lives.

Big Science Idea: Nature has evolved varied mechanisms to allow more beneficial genetic variation within a population and individuals, and populations with this variation are better equipped to survive and thrive in ever-changing environments.

Essential NGSS Standards:	Targeted CCSS ELA and ELD Standards
Make and defend a claim based on evidence that inheritable genetic variations may result from (1) new genetic combinations through meiosis, (2) viable errors occurring during replication, and/or (3) mutation caused by environmental factors. (HS-LS3-2)	**Interpretive:** Determine the central ideas or conclusions of a text; trace the text's explanation or depiction of a complex process, phenomenon, or concept; provide an accurate summary of the text.

Collaborative:

Adapting language choices to various contexts (based on task, purpose, audience, and text type).

Productive:

Write arguments focused on discipline-specific content. Specifically, develop claim(s) and counterclaims fairly, supplying data and evidence for each while pointing out the strengths and limitations of both claim(s) and counterclaims in a discipline-appropriate form and in a manner that anticipates the audience's knowledge level and concerns.

Anchor Phenomenon:	Anchor Text:
Whiptail lizards asexually reproduce without males in their population while still producing some level of genetic variation. Thus, are males needed in a population? Why or why not?	Yong, E. (2010). Extra chromosomes allow all-female lizards to reproduce without males. *Discover.* https://www .discovermagazine.com/planet-earth/extra -chromosomes-allow-all-female-lizards-to -reproduce-without-males

Based on the original article:

Lutes, A., Neaves, W., Baumann, D., Wiegraebe, W., & Baumann, P. (2010). Sister chromosome pairing maintains heterozygosity in parthenogenetic lizards. *Nature.* DOI: 10.1038/nature08818

Culminating Assessment Activity: Students write an evidence-based argument with the support of a claim–evidence–reasoning graphic organizer that responds to the following question: Which type of reproduction (sexual or asexual) allows a species/population to survive and thrive best over time? Why?

In their writing, students use evidence and reasoning from whiptail lizard close-read text, video resources discussing asexual versus sexual reproduction, meiosis modeling activities, paired discussion protocol, shared thinking with peers, and other resources when writing the evidence-based argument. (See Student Resources below.)

Targeted Language Demands

- **Language functions:** Argue with evidence.
- **Vocabulary:** asexual reproduction, sexual reproduction, parthenogensis, variation, meiosis, random segregation, haploid (N), diploid (2N), claim, counterclaim, evidence, reasoning.
- **Syntax:** Phrases that transition between and coordinate claim–evidence–reasoning (e.g., my evidence is, for example, in conclusion) See graphic organizer below.
- **Discourse:** Claim–evidence–reasoning structure. See graphic organizer below.

Student Resource: Claim-Evidence-Reasoning Argument Graphic Organizer

YOUR HYPOTHESIS/CLAIM	
(an explanation of a situation that is testable, verifiable, and refutable)	
EVIDENCE #1	Provide evidence that supports your claim.
According to . . .	
My evidence is . . .	*(Evidence is information from trusted sources such as texts, collaborative conversations, analyzed data from carefully performed scientific tests, etc., that supports your explanation.)*
Additional evidence is . . .	
Furthermore . . .	
For example,	
For instance,	
REASONING #1	Write a statement that explains *how/why* your evidence supports your claim.
Based on the evidence provided . . .	
This evidence supports my claim because . . .	
This confirms that . . . because . . .	
Therefore . . .	
In conclusion . . .	
EVIDENCE #2	Provide evidence that supports your claim.
According to . . .	
My evidence is . . .	
Additional evidence is . . .	
Furthermore . . .	
For example,	
For instance,	
REASONING #2	Write a statement that explains *why/how* your evidence supports your claim.
Based on the evidence provided . . .	
This evidence supports my claim because . . .	
This confirms that . . . because . . .	
Therefore . . .	
In conclusion . . .	

EVIDENCE #3	Provide evidence that supports your claim.
According to . . .	
My evidence is . . .	
Additional evidence is . . .	
Furthermore . . .	
For example,	
For instance,	

REASONING #3	Write a statement that explains *why/how* your evidence supports your claim.
Based on the evidence provided . . .	
This evidence supports my claim because . . .	
This confirms that . . . because . . .	
Therefore . . .	
In conclusion . . .	

Student Resource: Questions on the Reading "Extra Chromosomes Allow All-Female Lizards to Reproduce Without Males"

Directions

- Respond to each question below in two or more complete sentences that use <u>evidence from the text</u> as well as <u>your own thoughts and opinions</u>.
- Please include the question, your response, and the paragraph number where you can find the evidence for your answer.

Level 1 Questions (factual information; summarizes basic ideas from the text; answers found directly in text):

1. What are the benefits of sex?
2. How are sperm and egg cells produced during "normal" meiosis?
3. How much DNA/chromosomes do normal egg and sperm have?

Level 2 Questions (infer, interpret, or analyze based on the text; "how" or "why" questions that address what the text suggests rather than what it says):

4. How is whiptail meiosis different from "normal" meiosis?
5. Do whiptails reproduce sexually or asexually? Why?
6. If a whiptail lizard gives birth to a clutch of eggs, will the siblings be genetically identical? Why? Is this the same or different from a different lizard that reproduces sexually? Why?

Level 3 Questions (go beyond the text; reading asks to explore larger issues; use outside knowledge and experience to respond):

   7. Are males needed in a population? Why or why not?
   8. Which type of reproduction (sexual or asexual) allows a species/population to survive and thrive best over time? Why?

Student Resource: Paired/Group Productive Discussion

Directions

   1. Partner A *shares their claim, evidence, and reasoning* for their assigned question with Partner B.
   2. Partner B listens and takes notes on their partner's ideas throughout the discussion. Partner B *asks two or more questions from the "Nine Talk Moves"* (https://inquiryproject.terc.edu/prof_dev/Goals_and_Moves.cfm.html) to clarify Partner A's ideas and deepen their thinking.
   3. Partner B *shares their claim, evidence, and reasoning* for their assigned question with Partner A.
   4. Partner A listens and takes notes on their partner's ideas throughout the discussion. Partner A *asks two or more questions from the "Nine Talk Moves"* to clarify Partner B's ideas and deepen their thinking.

PARTNER NOTES:	PARTNER NOTES:
PARTNER A'S NAME:_____	PARTNER B'S NAME:_____
Paraphrase or summarize your partner's claim and thinking below:	*Paraphrase or summarize your partner's claim and thinking below:*
Additional ideas/comments after clarifying/deepening questions:	*Additional ideas/comments after clarifying/deepening questions:*

   5. Both partners **create a group claim** *together* based on their discussion and support with two or more pieces of evidence and reasoning from the text, the video, and/or their discussion. *Write on a separate piece of paper!*

UNIT III—HIGH SCHOOL CHEMISTRY: REPRESENTING THE FORMATION OF BREAD THROUGH A CHEMICAL EQUATION

Contribution by Anthony Bardessono

Unit Frame

Essential Student and Family Assets

- Personal interests: Some students have expressed specific interest or knowledge in subjects related to chemistry (cars/mechanics, baking, agriculture, working in service/tourist day jobs)
- Linguistic repertoire: The majority of students are emergent bilinguals. All but two have been redesignated as English proficient. These students are comfortable taking notes during class and employing visual representations of concepts in learning and expressing understanding. From previous work, I know they have strong artistic skill and enjoy projects that employ them.
- Home/family/community assets: For the unit specifically, all students will have some experience with bread, many families still make their own, and each culture has unique varieties. When I had previously brought up the chemistry of bread, some students expressed knowledge of the fermentation process. The high school is located in the center of a viticultural area, and some students have jobs in the service or tourism subindustries; many other students also had tacit knowledge of how wine was made, but many had not heard the term *fermentation* before or connected it to a chemical process as we had been discussing in class.

Big Science Idea: We can predict and measure how substances will change when they interact during a chemical reaction.

Essential NGSS Standards:	Targeted CCSS ELA and ELD Standards
HS-PS1-7. Use mathematical representations to support the claim that atoms, and therefore mass, are conserved during a chemical reaction.	**Interpretive:** Closely reading literary and informational texts and viewing multimedia to determine how meaning is conveyed explicitly and implicitly through language.
HS-PS1-2. Construct and revise an explanation for the outcome of a simple chemical reaction based on the outermost electron states of atoms, trends in the periodic table, and knowledge of the patterns of chemical properties.	**Collaborative:** Exchanging information and ideas with others through oral collaborative discussions on a range of social and academic topics. **Productive:** Justifying own arguments and evaluating others' arguments in writing.

(continued)

Anchor Phenomenon: Dough rising and transforming into bread.	**Anchor Text:** "Bread Science 101." *Science of Cooking.* https://www.exploratorium.edu/cooking/bread/bread_science.html

Culminating Assessment Activity: Throughout the unit students will be creating and iterating on a visual model (on poster paper) of dough rising and baking into bread. They will develop an initial model based on observations of the anchor phenomenon (the dough rising/baking), and return to and revise their models multiple times during the unit by folding in new information learned and responding to instructor and peer feedback. The following "Modeling Checklist" will guide ongoing development of the model. While students will be working in pairs on the model, each person will be assigned and responsible for half of the modeling checklist.

- *Does the model represent the phenomenon of dough rising (at the molecular and the observable levels)?*
- *Does the model incorporate appropriate scientific principles learned during class to explain how dough rises?*
- *Does the model use chemical formulas and balanced chemical equations to mathematically represent how dough rises?*
- *Does the model cite and use sources from research to support the explanation?*
- *Has the model been revised and improved from the prior version?*

The final version and evidence of change from previous version(s) of their model will be used as the unit's culminating assessment. They will also be asked to orally explain their model as part of a class "poster symposium" where other teachers, administrators, and community members are invited. They are also given the option to produce the model digitally.

Targeted Language Demands

- **Language functions:** Represent and explain a chemical process through a scientific model.
- **Vocabulary:** reactant, product, conservation, mass, energy, balance.
- **Syntax:** Appropriate chemical formula, labeling of the model.
- **Discourse:** Claim–evidence–reasoning structure.

Student Resource: Research Guide

- As you read, find and write down one other chemical reaction that happens in bread.
- What does the reaction above do as the dough is rising or baking?
- What are the chemical formulas for the reactants and products in the reaction?

Reactants:	Products:

- Now we want to make a chemical equation of our reaction and balance it:

$$\rightarrow$$

- What type of reaction is it? (Synthesis, decomposition, combustion, single-/double-replacement, or none of these)
- Cite your source(s) here. You can use Google to reach a website, but Google is *not* a source!:
  - » Website name:
  - » Author/editor (if any are listed):
  - » Publishing organization:

Blank Forms for Unit and Lesson Planning

PLANNING TOOL 1: KNOWING YOUR EMERGENT BILINGUALS AND THEIR FAMILIES

Guiding Questions

- How can you gather expanded relevant knowledge about emergent bilinguals and their families?
- How can you analyze and leverage this knowledge when planning your unit?

Part 1: Gather Expanded Relevant Knowledge (2–4 weeks before the unit begins)

Which methods will you use to gather knowledge about (1) personal interests and out-of-school experiences, (2) home/family/community funds of knowledge, and (3) ways to communicate in English and in languages other than English that are *relevant* to the science unit?

☐

☐

☐

Part 2: Analyze Knowledge Gathered

What have you learned about students and families that is relevant to this science unit?

- Personal interests and out-of-school experiences:

- Home/family/community funds of knowledge:

- Ways to communicate in English and in the home language:
  » Primary and home language context
  » Oral language
  » Written language

Part 3: Decide How to Leverage What You Learned About Students and Families Into a Unit

- What are the most essential student and family assets that connect to the big idea or learning task of the unit?

- How can they be used to frame the unit?

PLANNING TOOL 2: UNPACKING THE NEXT GENERATION SCIENCE STANDARDS AND CURRICULAR RESOURCES

Guiding Questions
• What are the essential NGSS performance expectations?
• How can you unpack standards and curricular resources to arrive at a big science idea statement, key vocabulary, and language function for the unit?

Unit Title:

Part 1: Essential NGSS Performance Expectations
Essential Performance Expectations of the Unit (3–5 max)

Focal **Disciplinary Core Idea(s) (DCI)**	

Focal **Science and Engineering Practice(s) (SEP)**	

Focal **Crosscutting Concept(s) (CCC)**	

Part 2: Unpacking Standards and Curricular Resources to Arrive at a Big Science Idea Statement, Key Vocabulary, and Focal Language Functions

Resources Reviewed

Key Concepts and the Big Science Idea

Key Supporting Vocabulary	**Core Idea-Related**	**SEP-Related**	**CCC-Related**

Focal Language Functions

PLANNING TOOL 3: WEAVING TOGETHER ELA/ ELD STANDARDS AND NGSS

Guiding Questions

- What collaborative, interpretive, and productive ELA/ELD standards will best support science learning and language/literacy development throughout the unit?
- What additional opportunity is there to support how English works?

Collaborative

Interpretive

Productive

Learning About How English Works

PLANNING TOOL 4: FRAMING THE UNIT THROUGH RICH AND RELEVANT PHENOMENA, TEXTS, AND ASSESSMENT

Guiding Questions

- What natural phenomenon or real-world problem would make the big idea rich and relevant?
- What texts could students be supported in interpreting throughout the unit?
- What culminating assessment activity could allow students to produce and share a text that explains the anchor phenomenon?

Big Idea

Anchor Phenomenon   Anchor Text   Culminating Assessment

PLANNING TOOL 5: CONCEPTUAL AND LINGUISTIC PROGRESSION OF LEARNING

Guiding Questions

- How can we generate a set of 3–5 checkpoints so that students are supported in progressing through the central science concepts and vocabulary en route to making sense of the big idea?
- How can we generate a set of 3–5 checkpoints so that students are supported in progressing through the interpretive, collaborative, and productive language en route to making sense of the big idea?

	Checkpoint #1	Checkpoint #2	Checkpoint #3	Endpoint
Concept				
Relevant Vocabulary (be able to *use*, not just define, these)				
Interpretive Language Progression				
Collaborative Language Progression				
Productive Language Progression				

PLANNING TOOL 6: ANALYZING THE ANCHOR TEXT

Guiding Question
How can we analyze the anchor text as a reader and as a teacher to better support interpretive language for students?

Title of Text:

Author/Composer:

Source:

Topic of Lesson:

Part 1: Analyze the text <u>as a reader</u>.

Closely read the text as a reader. As you read, complete and consider the following:

Mark significant words, notes, symbols, phrases, measures, and passages.

Reread the marked and annotated sections above and choose:

- the *three* passages/quotations/sections you believe will be the most significant for this lesson.
- the *four* most important words/phrases that will be the most significant for this lesson.

Add one more element of the text that resonates for you.

As a specialist in your discipline, what do *you* take away from this text?

What would you want *your students* to take away from this text?

Part 2: Analyze the text <u>in preparation for instruction</u>.

What **key concepts** in the text are *most important*?

What **background knowledge** would students *need* before engaging in the text?

Use of **multiple voices** (if needed in the text)

Use of **figurative language**
*Are there analogies, metaphors, or figurative or idiomatic expressions in the text?
Is any of the language potentially misleading?*

Purpose and **audience**
*What is the author's purpose? (To express, reflect, inquire, explore, inform,
explain, analyze, interpret, persuade, evaluate, judge, propose a solution, or seek
common ground?)*

*What kinds of readers does the author seem to anticipate? (What in the text tells
you that?) Is the text associated with a company, brand, organization, group, or
discipline?*

Part 3: Develop close-reading inquiry questions for the text.
*Remember to ask students to use <u>evidence from the text</u> in their responses and
the paragraph number(s) where evidence is found in the text.*

Level 1: questions ask for direct information from the text

Level 2: questions ask the student to infer, interpret, and analyze based on the text

Level 3: questions explore larger issues using/connecting outside knowledge and
experience to ideas from the text

PLANNING TOOL 7: UNPACKING THE UNIT'S SCIENCE-SPECIFIC LANGUAGE DEMANDS

Guiding Question

What science-specific language demands will be necessary during the unit for students to interpret, collaborate, and produce?

Language Function

(Active use of language for a
particular purpose and
audience, i.e., what you *do*
with language)

Discourse

(How scientists or engineers
talk, write, and reason about,
for a particular purpose and
audience, i.e., *tools* that
convey *meaning*)

Syntax

(How we organize words or
symbols in science to convey
meaning, i.e., tools that
organize)

UNIT FRAME

Essential Student Assets

- Personal interests:
- Linguistic repertoire:
- Home/family/community assets:
- Out-of-school science experiences:

Big Science Idea:

Essential NGSS Standards: **Targeted CCSS ELA and ELD Standards**

Interpretive:

Collaborative:

Productive:

Anchor Phenomenon: **Anchor Text:**

Culminating Assessment Activity:

Targeted Language Demands

- Language functions:
- Vocabulary:
- Syntax:
- Discourse:

LESSON PLAN

Central Focus
NGSS PE: ELA/ELD Standards:

Focal Science or Engineering Practice(s):	Focal Language Function(s):

Integrated Learning Objective:

Sequence of Opportunities for Learning Science and Using Language

Language and Literacy Supports		
Interpretive (making sense of oral and/or written language)	Collaborative (communicating with others to make meaning)	Productive (communicating meaning through oral and/or written language)

References

Ash, D. (2004). Reflective scientific sense making dialogue in two languages: The science in the dialogue and the dialogue in the science. *Science Education, 88*(6), 855–884.

August, D., McCardle, P., & Shanahan, T. (2014). Developing literacy in English language learners: Findings from a review of the experimental research. *School Psychology Review, 43*(4), 490–498.

Barton, A. C., & Tan, E. (2009). Funds of knowledge and discourses and hybrid space. *Journal of Research in Science Teaching, 46*(1), 50–73.

Banilower, E. R., Smith, P. S., Malzahn, K. A., Plumley, C. L., Gordon, E. M., & Hayes, M. L. (2018). *Report of the 2018 NSSME+*. Horizon Research.

Brown, B. A., & Ryoo, K. (2008). Teaching science as a language: A "content-first" approach to science teaching. *Journal of Research in Science Teaching, 45*(5), 529–553.

Bunch, G. C. (2013). Pedagogical language knowledge preparing mainstream teachers for English learners in the new standards era. *Review of Research in Education, 37*(1), 298–341.

Buxton, C. A., & Caswell, L. (2020). Next generation sheltered instruction to support multilingual learners in secondary science classrooms. *Science Education, 104*(3), 555–580.

California Department of Education. (2015). *California English language arts/English language development framework for California Public Schools: Kindergarten through grade twelve*. http://www.cde.ca.gov/ci/rl/cf/elaeldfrmwrksbeadopted.asp

Council of Chief State School Officers. (2013). *English language proficiency (ELP) standards*. http://www.elpa21.org/sites/default/files/Final%204_30%20ELPA21%20Standards_1.pdf

Crawford, J., & Adelman Reyes, S. (2015). *The trouble with SIOP®: How a behaviorist framework, flawed research, and clever marketing have come to define—and diminish—sheltered instruction for English language learners: Featuring an alternative approach to sheltered instruction and a sample unit applying that framework*. Institute for Language & Education Policy.

Cummins, J. (1980). The cross-lingual dimensions of language proficiency: Implications for bilingual education and the optimal age issue. *TESOL Quarterly, 14*(2), 175–187. https://doi.org/10.2307/3586312

Davis, E. A., & Krajcik, J. S. (2005). Designing educative curriculum materials to promote teacher learning. *Educational Researcher, 34*(3), 3–14.

DePaoli, J. L., Balfanz, R., Bridgeland, J., Atwell, M., & Ingram, E. S. (2017). *Building a grad nation: Progress and challenge in raising high school graduation rates. Annual update 2017*. Civic Enterprises.

Echevarria, J., Vogt, M., & Short, D. (2008). *Making content comprehensible for English learners: The SIOP model* (3rd ed.). Allyn & Bacon.

Faltis, C., & Ramirez-Marin, F. (2015). Secondary bilingual education: Cutting the gordian knot. In W. Wright, S. Boun, & O. Garcia (Eds.), *Handbook of bilingual and multilingual education* (pp. 336–353). Wiley-Blackwell.

Faltis, C. J., & Valdés, G. (2016). Preparing teachers for teaching in and advocating for linguistically diverse classrooms: A vade mecum for teacher educators. In D. Gitomer & C. Bell (Eds.), *Handbook of research on teaching* (5th ed., pp. 549–592). American Educational Research Association.

García, O. (2009). Education, multilingualism and translanguaging in the 21st century. In T. Skutnabb-Kangas, R. Phillipson, A. K. Mohanty, & M. Panda (Eds.), *Social justice through multilingual education* (pp. 140–158). Multilingual Matters.

Goldenberg, C. (2013). Unlocking the research on English learners: What we know—and don't yet know—about effective instruction. *American Educator, 37*(2), 4–11.

González, N., Moll, L. C., & Amanti, C. (2005). *Funds of knowledge: Theorizing practices in households, communities and classrooms*. Lawrence Erlbaum Associates.

Grossman, P., Hammerness, K., & McDonald, M. (2009). Redefining teaching, re-imagining teacher education. *Teachers and Teaching: Theory and Practice, 15*(2), 273–289.

Guilford, J., Bustamante, A., Mackura, K., Hirsch, S., Lyon, E., & Estrada, K. (2017). Text savvy. *The Science Teacher, 84*(1), 49–56.

Hakuta, K., Santos, M., & Fang, Z. (2013). Challenges and opportunities for language learning in the context of the CCSS and the NGSS. *Journal of Adolescent & Adult Literacy, 56*(6), 451–454.

Hawkins, M. (2004). Researching English language and literacy development in schools. *Educational Researcher 33*(3), 14–25.

Kibler, A., Valdés, G., & Walqui, A. (2014). What does standards-based educational reform mean for English language learner populations in primary and secondary schools? *TESOL Quarterly, 48*(3), 433–453. https://doi.org/10.1002/tesq.183

Kavanagh, S. S., & Rainey, E. C. (2017). Learning to support adolescent literacy: Teacher educator pedagogy and novice teacher take up in secondary English language arts teacher preparation. *American Educational Research Journal, 54*(5), 904–937.

Krashen, S. D. (1981). *Second language acquisition and second language learning*. Pergamon Press.

Lara-Alecio, R., Tong, F., Irby, B. J., Guerrero, C., Huerta, M., & Fan, Y. (2012). The effect of an instructional intervention on middle school English learners' science and English reading achievement. *Journal of Research in Science Teaching, 49*(8), 987–1011.

Lee, O., Quinn, H., & Valdés, G. (2013). Science and language for English language learners in relation to Next Generation Science Standards and with implications for Common Core State Standards for English language arts and mathematics. *Educational Researcher, 42*(4), 223–233. https://doi.org/10.3102/0013189X13480524

Lee, O., & Stephens, A. (2020). English learners in STEM subjects: Contemporary views on STEM subjects and language with English learners. *Educational Researcher, 49*(6), 426–432.

Lemmi, C., Brown, B. A., Wild, A., Zummo, L., & Sedlacek, Q. (2019). Language ideologies in science education. *Science Education, 103*(4), 854–874.

Llosa, L., Lee, O., Jiang, F., Haas, A., O'Connor, C., Van Booven, C. D., & Kieffer, M. J. (2016). Impact of a large-scale science intervention focused on English language learners. *American Educational Research Journal, 53*(2), 395–424.

Lyon, E. G., Stoddart, T., Bunch, G. C., Tolbert, S., Salinas, I., & Solis, J. (2018). Improving the preparation of novice secondary science teachers for English learners: A proof of concept study. *Science Education, 102*(6), 1288–1318.

Lyon, E. G., Tolbert, S., Solís, J., Stoddart, T., & Bunch, G. C. (2016). *Secondary science teaching for English learners: Developing supportive and responsive learning contexts for sense-making and language development.* Rowman & Littlefield Publishers.

Michaels, S., & O'Connor, C. (2012). *Talk science primer.* TERC.

McDonald, M., Kazemi, E., & Kavanagh, S. S. (2013). Core practices and pedagogies of teacher education: A call for a common language and collective activity. *Journal of Teacher Education, 64*(5), 378–386.

National Academies of Sciences, Engineering, and Medicine. (2018). *English learners in STEM subjects: Transforming classrooms, schools, and lives.* National Academies Press.

National Center for Education Statistics. (2016). *NAEP - 2015 science assessment.* The Nation's Report Card. https://www.nationsreportcard.gov/science_2015/#?grade=4

National Governors Association Center for Best Practices & Council of Chief State School Officers. (2010). *Common Core State Standards for English Language Arts.*

National Research Council (NRC). (1996). *National science education standards.* The National Academies Press.

National Research Council (NRC). (2001). *Knowing what students know: The science and design of educational assessment.* The National Academies Press. https://doi.org/10.17226/10019

National Research Council (NRC). (2012). *A framework for K–2 science education: Practices, crosscutting concepts, and core ideas.* The National Academies Press.

NGSS Lead States. (2013a). *Next Generation Science Standards: For states, by states.* The National Academies Press.

NGSS Lead States. (2013b). *Next Generation Science Standards: For states, by states. Appendix D: All standards, all students.* The National Academies Press.

Paek, S., & Fulton, L. (2021). Digital science notebooks: A tool for supporting scientific literacy at the elementary level. *TechTrends, 65*(3), 359–370.

Quinn, H., Lee, O., & Valdés, G. (2012). Language demands and opportunities in relation to Next Generation Science Standards for English language learners: What teachers need to know. In *Commissioned papers on language and literacy issues in the Common Core State Standards and Next Generation Science Standards* (pp. 32–43). Understanding Language.

Rodriguez, A. J. (2015). What about a dimension of engagement, equity, and diversity practices? A critique of the Next Generation Science Standards. *Journal of Research in Science Teaching, 52*(7), 1031–1051.

Salloum, S., Siry, C., & Espinet, M. (2020). Examining the complexities of science education in multilingual contexts: Highlighting international perspectives. *International Journal of Science Education, 42*(14), 2285–2289.

Shanahan, T., & Shanahan, C. (2012). What is disciplinary literacy and why does it matter? *Topics in Language Disorders, 32*(1), 7–18.

Shepard, L. A. (2013). Foreword. In J. H. McMillan (Ed), *SAGE handbook of research on classroom assessment* (pp. xix–xxii). SAGE.

Stroupe, D. (2015). Describing "science practice" in learning settings. *Science Education, 99*(6), 1033–1040.

Suárez, E. (2020). "Estoy Explorando Science": Emergent bilingual students problematizing electrical phenomena through translanguaging. *Science Education, 104*(5), 791–826.

Thompson, J., Hagenah, S., Kang, H., Stroupe, D., Braaten, M., Colley, C., & Windschitl, M. (2016). Rigor and responsiveness in classroom activity. *Teachers College Record, 118*(5), 1–58. https://doi.org/10.1177/016146811611800506

Tolbert, S., & Knox, C. (2016). 'They might know a lot of things that I don't know': Investigating differences in preservice teachers' ideas about contextualizing science instruction in multilingual classrooms. *International Journal of Science Education, 38*(7), 1133–1149.

Walqui, A., & Bunch, G. C. (Eds.). (2019). *Amplifying the curriculum: Designing quality learning opportunities for English learners.* Teachers College Press.

Warren, B., Ballenger, C., Ogonowski, M., Rosebery, A. S., & Hudicourt-Barnes, J. (2001). Rethinking diversity in learning science: The logic of everyday sensemaking. *Journal of Research in Science Teaching, 38*(5), 529–552.

Windschitl, M., Thompson, J., & Braaten, M. (2018). *Ambitious science teaching.* Harvard Education Press.

Windschitl, M., Thompson, J., Braaten, M., & Stroupe, D. (2012). Proposing a core set of instructional practices and tools for teachers of science. *Science Education, 96*(5), 878–903.

Index

The letter *f* after a page number refers to a figure.

About the Authors

Edward G. Lyon is an associate professor of science education at Sonoma State University and a former high school biology, chemistry, and sheltered learning science teacher. He holds a PhD in science education from the University of California Santa Cruz. Lyon's scholarship investigates the role of teacher education in supporting how science teachers integrate language, literacy, and science in multilingual classrooms. Dr. Lyon directs and is principal investigator (PI) for the Biliteracy and Content Area Integrated Preparation (BCAIP) Project funded by the U.S. Department of Education. He formerly served as co-PI for the National Science Foundation–funded Secondary Science Teaching with English Language and Literacy Acquisition (SSTELLA) Project. Lyon has published in leading science education journals and cowrote the book *Secondary Science Teaching for English Learners: Developing Supportive and Responsive Learning Contexts for Sense-Making and Language Development.*

Kelly M. Mackura is a middle school science teacher in Santa Rosa, California. She has taught middle and high school biology, earth science, and sheltered learning for over 20 years. She also holds an administrative services credential and an MA in education from Sonoma State University. She recently served as a Teacher on Special Assignment (TOSA) for Santa Rosa City Schools to facilitate NGSS implementation. As a mentor of preservice science teachers, Mackura has acted as a lead teacher for multiple teacher education initiatives with Sonoma State. She developed components of the model unit discussed throughout this book as part of her culminating MA in education project to understand how to elicit students' funds of knowledge, including families, to inform language- and literacy-rich science instruction.